NUTRITION AND DIET RESEARCH PROGRESS

WHAT TO KNOW ABOUT CAROTENOIDS

NUTRITION AND DIET RESEARCH PROGRESS

Additional books and e-books in this series can be found on Nova's website under the Series tab.

NUTRITION AND DIET RESEARCH PROGRESS

WHAT TO KNOW ABOUT CAROTENOIDS

ROBERT M. ALBERT
EDITOR

Copyright © 2021 by Nova Science Publishers, Inc.

All rights reserved. No part of this book may be reproduced, stored in a retrieval system or transmitted in any form or by any means: electronic, electrostatic, magnetic, tape, mechanical photocopying, recording or otherwise without the written permission of the Publisher.

We have partnered with Copyright Clearance Center to make it easy for you to obtain permissions to reuse content from this publication. Simply navigate to this publication's page on Nova's website and locate the "Get Permission" button below the title description. This button is linked directly to the title's permission page on copyright.com. Alternatively, you can visit copyright.com and search by title, ISBN, or ISSN.

For further questions about using the service on copyright.com, please contact:
Copyright Clearance Center
Phone: +1-(978) 750-8400   Fax: +1-(978) 750-4470   E-mail: info@copyright.com.

NOTICE TO THE READER

The Publisher has taken reasonable care in the preparation of this book, but makes no expressed or implied warranty of any kind and assumes no responsibility for any errors or omissions. No liability is assumed for incidental or consequential damages in connection with or arising out of information contained in this book. The Publisher shall not be liable for any special, consequential, or exemplary damages resulting, in whole or in part, from the readers' use of, or reliance upon, this material. Any parts of this book based on government reports are so indicated and copyright is claimed for those parts to the extent applicable to compilations of such works.

Independent verification should be sought for any data, advice or recommendations contained in this book. In addition, no responsibility is assumed by the Publisher for any injury and/or damage to persons or property arising from any methods, products, instructions, ideas or otherwise contained in this publication.

This publication is designed to provide accurate and authoritative information with regard to the subject matter covered herein. It is sold with the clear understanding that the Publisher is not engaged in rendering legal or any other professional services. If legal or any other expert assistance is required, the services of a competent person should be sought. FROM A DECLARATION OF PARTICIPANTS JOINTLY ADOPTED BY A COMMITTEE OF THE AMERICAN BAR ASSOCIATION AND A COMMITTEE OF PUBLISHERS.

Additional color graphics may be available in the e-book version of this book.

Library of Congress Cataloging-in-Publication Data

ISBN: 978-1-68507-105-9

Published by Nova Science Publishers, Inc. † New York

CONTENTS

Preface   vii

Chapter 1   Astaxanthin as a Valuable Natural Carotenoid   **1**
*Farah Ayuni Mohd Hatta, Rashidi Othman,
Qurratu Aini Mat Ali, Razanah Ramya,
Nur Hanie Mohd Latif
and Wan Syibrah Hanisah Wan Sulaiman*

Chapter 2   Challenges in Oral Bioavailability of Lutein   **27**
*Ishani Bhat, Saisree Iyer, Vanessa Fernandes,
Divyashree M and Bangera Sheshappa Mamatha*

Chapter 3   Technological Approaches
to the Production of Yeast Carotenoids   **67**
*Lucielen Oliveira Santos,
Pedro Garcia Pereira Silva,
Daniel Prescendo Júnior,
Tabita Veiga Dias Rodrigues,
Erika Carvalho Teixeira
and Janaina Fernandes de Medeiros Burkert*

Chapter 4	Rosaceae Fruits as a Valuable Source of Carotenoids. A Review *T. Negreanu-Pirjol, D. R. Popoviciu and B. S. Negreanu-Pirjol*	**111**
Index		**159**

PREFACE

Carotenoids are the yellow, red, and orange pigments that are produced by various plants, algae, fungi, and bacteria, and can act as antioxidants in humans. In chapter one of this four-chapter monograph, the authors describe astaxanthin, a reddish carotenoid found in seafood with potent antioxidant characteristics. Chapter two compiles the challenges encountered in the oral bioavailability of lutein, a xanthophyll carotenoid. Chapter three reviews the main parameters of the production of yeast carotenoids and the downstream processes to obtain carotenogenic extracts. Chapter four summarizes known data on total carotenoid contents and chemical diversity of Rosaceae fruits.

Chapter 1 - Carotenoids can be obtained naturally from several sources like fruits, algae, plants, and crustaceans, contributing to varying red or yellow pigmentation. Carotenoids are acyclic substances having several transformations, leading to more than 800 compounds derived by the addition of oxygen-abundant functional groups or end group cyclisation. Hence, the polyene composition allows carotenoids to be potent free-radical scavengers. These compounds can capture single oxygen atoms or peroxyl radicals. Astaxanthin leads the global carotenoid market that is expected to reach $427.0 million in 2022 from $288.7 million in 2017, corresponding to an 8.1% CAGR. Astaxanthin itself is expected to have a global market value of $2.57 billion by 2025. Astaxanthin demand is rising due to increasing health awareness among consumers who demand products and applications, leading to a $2000 per kg market price. Moreover, astaxanthin possesses colour imparting properties that also increase market demand. Astaxanthin (3,3'-dihydroxy-,-carotene-4,4'-dione) is a xanthophyll carotenoid with lipophilic characteristics (C_{40}). This substance is reddish and found abundantly in crustaceans like crab, shrimp, lobster, salmon, and other seafood. Phytoplankton and specific microalgae are the significant consumers of astaxanthin, which builds up inside the bodies of these organisms. Astaxanthin is a shrimp by-product with potent antioxidant characteristics beneficial for humans since they provide anti-inflammatory and UV-light defence properties. Numerous studies indicate very high biological activity compared to other carotenoids; Astaxanthin is ten times more potent than lutein, canthaxanthin, β-carotene, and zeaxanthin, while it is 100 times more potent than α-tocopherol. Such characteristics are attributed to the specific molecular arrangement comprising keto moieties and hydroxyl presence on every ionone ring and long conjugated double bonds (13 conjugated double bonds) in the middle of the structure. Much research has been dedicated to assess Astaxanthin processing, functionality, and sources. Nevertheless, data concerning compound stability considering various environmental, storage, and carrier conditions is scarce.

Chapter 2 - Lutein is a xanthophyll carotenoid having a C40 isoprenoid (conjugated double bonds) backbone with oxygen-containing rings at both ends and is responsible for yellow-orange color in many fruits and vegetables. The conjugated double bonds confer lutein potent anti-oxidative property to manifest considerable biological activities that provide specific health benefits. Since lutein is synthesized only in photosynthetic organisms, it is categorized as an important dietary carotenoid to humans. On oral consumption of lutein, the effective concentration available at the target site depends on the amount consumed, absorbed, transported and metabolized within the host. Further, the bioactivity of lutein is determined by the compound's bioaccessibility (the amount of lutein liberated from the food matrix during digestion) and bioavailability (the amount of lutein that enters the site of function). Since lutein is light and heat-sensitive pigment, processing conditions such as cooking, frying, and baking on dietary sources can reduce its levels before consumption. Several intrinsic and extrinsic factors influence the oral bioavailability of lutein. The key factors that influence carotenoid absorption are jointly abbreviated as SLAMENGI. It includes *S*pecies of the carotenoid, *L*inkage of the molecules, *A*mount of carotenoid consumed, *M*atrix surrounding the carotenoid, *E*ffectors of absorption and bioconversion, *N*utrition status of the host, *G*enetic variations of the host, and mathematical *I*nteractions. Lutein is a fat-soluble phytochemical that has been associated with the absorption of other lipophilic molecules like dietary lipids. It is released from the food matrix when ingested orally due to mechanical disruption and gastrointestinal juices. However, the oral bioavailability of lutein depends on its solubilization, degradation and rate of absorption. Additionally, variations in the genes that encode proteins involved in the absorption, transportation and metabolization of lutein within the host also modulate the oral bioavailability of lutein. This chapter is henceforth a compilation of the challenges encountered in the oral bioavailability of lutein.

Chapter 3 - Carotenoid pigments, which represent the most widespread group of natural pigments, are yellow, orange and red lipophilic substances whose structures exhibit much diversity. They are synthesized by plants, algae and several microorganisms, such as bacteria, microalgae, filamentous fungi and yeasts. Despite their wide distribution in nature, industrial commercialization of these pigments is mainly obtained by chemical synthesis. Biotechnological synthesis is an interesting alternative since it is natural and commercially competitive, based on consumers' concern over excessive use of chemical additives in food. Therefore, improving efficiency of carotenoid biosynthesis during fermentation/cultivation is crucial. New strategies to obtain these biomolecules through the optimization of biotechnological processes which use yeasts and the development of new approaches to downstream steps have been studied to make them become economically viable and competitive in the industrial market. Thus, this chapter aims at reviewing the main parameters of the production of yeast carotenoids and the downstream processes to obtain carotenogenic extracts. As a result, several aspects, such as the medium composition (carbon and nitrogen concentration, alternative raw materials), technological parameters (agitation/aeration, pH, temperature, fed-batch process, magnetic field application, light irradiation), genetic engineering, knowledge about carotenoid recovery and potential applications, were considered.

Chapter 4 - Rosaceae is one of the largest families of flowering plants, comprising numerous species of culinary, medicinal and ornamental interest, featuring a wide variety of fruit types: pomes, drupes, polydrupes or polyachenes. This paper summarizes known data on total carotenoid contents and chemical diversity of Rosaceae fruits. Overall carotenoid content within this family is extremely variable, with some commonly grown fruits (apples, pears, cherries, strawberries) ranking among the lowest, while tropical *Rubus* species, followed by species in the *Sorbus*, *Cotoneaster*, *Pyracantha* or *Rosa* genera having remarkably high carotenoid contents. Among the known carotenoid compounds found in Rosaceae fruits are α-, β- and γ-carotene, β-cryptoxanthin, zeaxanthin, violaxanthin, rubixanthin, flavoxanthin, neoxanthin, lycoxanthin, capsanthin, xantophyll esters, lutein, lutein epoxide, lycopene, prolycopene, β-citraurin, capsorubin etc., with variations among different species. Also, carotenoid content shows significant variations between fruit skins and flesh, with skins having higher amounts.

In: What to Know about Carotenoids
Editor: Robert M. Albert
ISBN: 978-1-68507-105-9
© 2021 Nova Science Publishers, Inc.

Chapter 1

ASTAXANTHIN AS A VALUABLE NATURAL CAROTENOID

Farah Ayuni Mohd Hatta[1], Rashidi Othman[2,∗], Qurratu Aini Mat Ali[1], Razanah Ramya[3], Nur Hanie Mohd Latif[4] and Wan Syibrah Hanisah Wan Sulaiman[4]

[1]Institute of Islam Hadhari, The National University of Malaysia, Bangi, Selangor Darul Ehsan, Malaysia
[2]Herbarium Unit, Department of Landscape Architecture, Kulliyyah of Architecture and Environmental Design, International Islamic University Malaysia, Kuala Lumpur, Malaysia
[3]Institute of the Malay World and Civilization, The National University of Malaysia, Bangi, Selangor Darul Ehsan, Malaysia
[4]International Institute for Halal Research and Training, International Islamic University Malaysia, Kuala Lumpur, Malaysia

∗ Corresponding Author's E-mail: rashidi@iium.edu.my.

ABSTRACT

Carotenoids can be obtained naturally from several sources like fruits, algae, plants, and crustaceans, contributing to varying red or yellow pigmentation. Carotenoids are acyclic substances having several transformations, leading to more than 800 compounds derived by the addition of oxygen-abundant functional groups or end group cyclisation. Hence, the polyene composition allows carotenoids to be potent free-radical scavengers. These compounds can capture single oxygen atoms or peroxyl radicals. Astaxanthin leads the global carotenoid market that is expected to reach $427.0 million in 2022 from $288.7 million in 2017, corresponding to an 8.1% CAGR. Astaxanthin itself is expected to have a global market value of $2.57 billion by 2025. Astaxanthin demand is rising due to increasing health awareness among consumers who demand products and applications, leading to a $2000 per kg market price. Moreover, astaxanthin possesses colour imparting properties that also increase market demand. Astaxanthin (3,3'-dihydroxy-,-carotene-4,4'-dione) is a xanthophyll carotenoid with lipophilic characteristics (C_{40}). This substance is reddish and found abundantly in crustaceans like crab, shrimp, lobster, salmon, and other seafood. Phytoplankton and specific microalgae are the significant consumers of astaxanthin, which builds up inside the bodies of these organisms. Astaxanthin is a shrimp by-product with potent antioxidant characteristics beneficial for humans since they provide anti-inflammatory and UV-light defence properties. Numerous studies indicate very high biological activity compared to other carotenoids; Astaxanthin is ten times more potent than lutein, canthaxanthin, β-carotene, and zeaxanthin, while it is 100 times more potent than α-tocopherol. Such characteristics are attributed to the specific molecular arrangement comprising keto moieties and hydroxyl presence on every ionone ring and long conjugated double bonds (13 conjugated double bonds) in the middle of the structure. Much research has been dedicated to assess Astaxanthin processing, functionality, and sources. Nevertheless, data concerning compound stability considering various environmental, storage, and carrier conditions is scarce.

Keywords: carotenoids, astaxanthin, bioactivity, extraction, pigment, stability

INTRODUCTION

Living organisms synthesise natural colourants that can be divided into three categories: tetrapyrroles, flavonoids, and tetraterpenoids (Aberoumand 2011). Anthocyanin is a flavonoid found in several fruits, including berries; the compound provides red and purple shades to fruits. Tetrapyrroles contain chlorophyll that is regarded as the most critical element present in higher plants. Like chlorophyll, carotenoids similar to tetraterpenoids are typically present since they absorb sunlight to complete photosynthesis and reduce light-induced chlorophyll degradation. Fundamentally, several non-photosynthetic organisms such as fungus and all photosynthetic plants and organisms produce carotenoids that belong to the isoprenoid metabolite class (Rodríguez-Concepción et al., 2018). More than 1000 different carotenoids are discovered, and all comprise double-bonded polyene chains, classified as xanthophylls and carotenes (Yabuzaki, 2017; Hermanns et al., 2020).

Carotenoids possess distinct colours because of their structure. Plant-based carotenoids typically comprise C_{40} tetraterpenes (carotenoids like lycopene, lutein, carotene, zeaxanthin, and cryptoxanthin) that facilitate plant growth (Rodríguez-Concepción et al., 2018; Wurtzel, 2019; Sun and Li, 2020; Zheng et al., 2020). Plant sustainability is maintained by green tissues that comprise carotenoid metabolites like zeaxanthin, violaxanthin, lutein, and β-carotene. Additionally, photosynthesis depends on carotenoids as vital pigments that help absorb 400–550 nm wavelength radiation that chlorophylls cannot use (Hermanns et al., 2020). In fact, human health and nutrition also depends on carotenoids as vital metabolites (Eggersdorfer and Wyss, 2018; Rodríguez- Concepción et al., 2018). Meléndez-Martínez (2019) indicated that carotenoids facilitate the production of vitamin A, micronutrients; they also regulate immune response, influence vision and reproductive processes.

Furthermore, carotenoids comprise an efficacious mix of bioactive substances like antioxidants (β-carotene and lycopene) that can reduce the incidence of chronic ailments like cancer and cardiovascular diseases (Bahukhanid et al., 2018; Hermanns et al., 2020). Epidemiological research suggests carotenoids possess numerous nutrition and health benefits for human beings. Outcomes also suggest that carotenoid inclusion in the diet through vegetables and fruits is positively associated with a reduced incidence of chronic conditions (Bohn, 2016).

α-carotene, β-carotene, and lycopene are pure hydrocarbon carotenoids; on the other hand, xanthophylls have oxygen-containing functional groups such as astaxanthin, zeaxanthin, lutein, and β-cryptoxanthin (Rivera, Vilaró & Canela, 2011; Fortes, 2006; Rodriguez-Amaya, 2001). β-carotene is among the famous food carotenoids; some foods have both α-carotene and β-carotene. α-carotene is present in pumpkin and carrot, while β-carotene is present in apricot, carrot, and mango. Green vegetables and orange or yellow flowers and fruits comprise lutein, which is a dihydroxy variant of β–carotene. Also, several red vegetables and fruits like grape, tomato, watermelon, and pink guava contain carotenoid lycopene (Fortes, 2006; Rodriguez-Amaya, 2001).

On the other hand, astaxanthin is a vital carotenoid present in marine creatures like crab, lobster, salmon, shrimp, and other microorganisms (Ushakumari and Ramanujan, 2012). Several carotenoids such as bixin might be obtained from annatto; saffron contains crocin (Rodriguez-Amaya 2001). Bixin is used for obtaining red pigment utilised for cosmetic, textile, food, and pharmaceutical applications (Santos, Albuquerque & Meireles, 2011). Primary carotenoids present in food are affected by locality, genetics, handling mechanisms, and seasonality (Arvayo-Enríquez, Mondaca-Fernández, Gortáres-Moroyoqui, Lopez-Cervantes, & Rodríguez-Ramírez, 2013; Fortes, 2006; Othman, 2009).

BCC Research Report (McWilliams, 2018) suggests that global carotenoid market value is expected to rise to $2.0 billion in 2022 from $1.5 billion in 2017, corresponding to a 5.7% CAGR during 2017-2022. Geriatric population increase specific to disease prevention and healthcare is a crucial aspect influencing carotenoid demand. Additionally, the need for natural colourants augments demand. Research and Development (R&D) for discovering natural carotenoids having significant value is forecast duration prospect (McWilliams, 2018). The global carotenoid market forecast provided by BCC Research; data is categorised for the forecast duration 2017 – 2024; indicates that astaxanthin is the most demanded carotenoid with a $288.7 million value in 2017. It is forecasted to witness an 8.1% CAGR and reach $427.0 million by 2022. By 2025, astaxanthin is expected to command a $2.57 billion market value globally.

Chemical Structure and Bioactive Compound Properties of Carotenoids

Acyclic C_{40} isoprenoid lycopene is the precursor required for carotenoid production; the former is considered a tetraterpene (Arvayo-Enríquez et al., 2013). Typically, most carotenoids are soluble in organic solvents like chloroform, ethyl ether, alcohol, acetone, and ethyl acetate because of their lipophilicity; consequently, they are water-insoluble. Hence, hexane, toluene, or petroleum ether are ideal for dissolving carotenes; xanthophylls are best dissolved by ethanol and methanol (Rodriguez-Amaya 2001). However, most food carotenoids are present in the form of all-*trans* polytenes comprising 5-carbon isoprenoid structures. Nevertheless, carotene processing, especially thermal, leads to a minute fraction of cis-isomer formation (Arvayo-Enríquez et al., 2013; Borel, 2003).

Phytoene-lycopene conversion typically comprises the addition of a single double-bond to the molecule. Lycopene contains between three and thirteen double bonds. Lycopene formation triggers end group enzymatic cyclisation process, yielding γ-carotene (one beta ring) and β-carotene (two beta rings). Source carotenoid concentration limits the biosynthetic process (Fortes, 2006). For instance, the extreme concentration of lycopene found in red tomatoes leads to the infeasibility of conversion to β-carotene because of limited enzyme activity. McClements, Decker, Park & Weiss (2009) suggested that specific cases exhibited the production of 6 carbon ring structures on one or both molecular endpoints. Farre et al., (2015) asserted that all carotenoids exhibited a polyisoprenoid formation, with the centre having long conjugated double bonds. The acyclic structure can convert and produce an extensive set of over 800 substances by end group cyclisation or the addition of oxygen-rich functional groups (Britton, Liaaen Jensen, and Pfander, 2004).

Carotenoids have a polyene structure that allows the substances to absorb free radicals by trapping peroxyl radicals or quenching singlet oxygen species. Fundamentally, compounds exhibit antioxidant characteristics in proportion to the presence of conjugated chains. Also, end group (like hydroxyl and carbonyl) polarity caused by β-ionone rings leads to significantly increase antioxidant activity. Extracts produced from natural sources typically comprise bioactive substances possessing antimicrobial, antiviral, antifungal, antibacterial, anti-inflammatory, antitumor, and antioxidant characteristics (da Silva, Rocha-Santos, and Duarte, 2016). β–cryptoxanthin, β-carotene, and γ-carotene are considered pro-vitamin A, unlike xanthophylls and lycopene (Arvayo-Enríquez et al., 2013). Carotenoids can potentially reduce cell membrane injury, regulate cellular immune response, arrest the growth of tumour cells, and attach to free-radical triggered receptors (Fortes, 2006). Along the same lines, carotenoid annatto is demonstrated to possess potent antioxidant characteristics, offering protection from sunlight and free radicals (Santos et al., 2011). Additionally, several epidemiological studies show that dietary carotenoid consumption might reduce cancer risk, indicating the criticality of antioxidant characteristics for cancer prevention. Nevertheless, data about β–cryptoxanthin, β-carotene, lutein, zeaxanthin, and α-carotene dosage as supplements against cancer remains inadequate; studies must address these issues in a balanced-diet context (Fortes, 2006). Prior research (Mohd Hatta and Othman, 2020) lists carotenoid sources and constituent bioactive substances.

Astaxanthin

Astaxanthin (3,3'-dihydroxy-β, β-carotene-4,4'-dione) is a C_{40} lipophilic carotenoid belonging to the xanthophyll category (Rodrigo-Baños et al., 2015). Figure 1.1 depicts the molecular structure comprising a polyene chain attached to two terminal rings and asymmetric carbon atoms on position 3. Astaxanthin has $C_{40}H_{52}O_4$ molecular formula corresponding to a 596.85 g/mol molar mass (Ambati et al., 2014). It is a red coloured lipophilic carotenoid that is abundant typically in crustaceans like crab, salmon, shrimp, lobster or other seafood (Radzali et al., 2016; Rodriguez-Amaya, 2001; Sui et al., 2015; Mezzomo and Ferreira, 2016). Some phytoplankton and microalgae contain astaxanthin because they are primary consumers of this substance (Senthamil and Kumaresan, 2015).

Astaxanthin obtained from shrimp discard (by-product) is considered to have significant antioxidant characteristics that benefit human health by promoting anti-inflammatory effects and UV-light protection (Ushakumari & Ramanujan, 2012; Sui et al., 2015). Astaxanthin is not a prerequisite for vitamin A formation; however, numerous studies suggest that it has better biological characteristics compared to other carotenoids (Ambati et al., 2014; Lin, Chen, Chen, Chen, & Ho, 2016; Mezzomo & Ferreira, 2016). The antioxidant effects of astaxanthin are ten times more potent than lutein, β-carotene, canthaxanthin, and zeaxanthin; moreover, its potency is hundred times that of α-tocopherol. Potency difference may be attributed to its unique molecular arrangement comprising keto moieties and hydroxyl groups on every ionone ring; moreover, the centre of the molecule comprises a long chain of thirteen conjugated double bonds (Delgado-Vargas and Paredes-Lopez, 2000). Previous studies have evaluated astaxanthin stability considering various storage and carrier scenarios (Ambati et al., 2014; Chen et al., 2007; Lin et al., 2016; Kittikaiwan et al., 2007; Gómez-Estaca et al., 2017; Sachindra and Mahendrakar, 2010).

Biotechnological advancement and credence of the remarkable biological characteristics of astaxanthin have provided a growth impetus for several commercial uses. The carotenoid finds extensive use as a supplement used for nutraceutical, food, and pharmaceutical applications (Ambati et al., 2014). Furthermore, the United States Food and Drug Administration (USFDA) has recognised the benefits of astaxanthin and approved it as a colouring agent (pigment) for animal feed. Hence, this action has prompted several researchers to enhance astaxanthin production using natural substances to offset the need for the synthetic version, thereby realising better market price. Grand View Research (2017) suggested that by 2025, astaxanthin is expected to have a global market value of $2.57 billion. Consumers are becoming increasingly health-conscious and demanding more astaxanthin products and use areas. As per Li, Zhu, Niu, Shen, and Wang (2011), the carotenoid might reach a market price of over $2000 per kg.

Astaxanthin Production from Shrimp Waste

World annual production of crustacean discard ranges between six to eight million tons (Yan and Chen, 2015). The statistical arm of the Food and Agriculture Organization (FAOSTAT) indicated that Europe alone produces crustacean waste of over 750,000 tons per annum (Gruber, 2013). Similarly, this figure stands at 1.5 million tonnes for Southeast Asia (FAO, 2014). Wang et al., (2007) suggested that crustacean harvest produces more than 50% waste that degrades over several years. Malaysia is witnessing an increase in shrimp aquaculture areas; moreover, the government is incentivising the industry considering the economic potential of shrimp aquaculture. Therefore, such waste is expected to rise significantly in Malaysia (Hashim and Kathamuthu, 2005). Crustaceans were valued at $57.1 billion for 7.9 million tonnes of which, the white leg shrimp *Litopenaeus vannamei* comprises over 53% of the crustacean production.

Nevertheless, rising shrimp aquaculture creates extensive ecological effects, including the massive shrimp waste produced after harvesting (Famino, Oduguwa, Onifade, and Olotunde, 2000). Shrimp processing produces waste from the head of the cephalothorax and the exoskeleton; these account for about 70% of the raw waste (Quan & Turner, 2009; Mezzomo, Martínez, Maraschin & Ferreira, 2013). Hence, researchers have suggested different substitutes to mitigate the problem and practice aquaculture sustainably. Shrimp waste is found to have usable carotenoid materials that might be usable for pharmaceutical, nutraceutical, and food applications (Mezzomo et al., 2013; Ambati et al., 2014; Sui et al., 2015).

Hence, carotenoids might be extracted from the waste cost-effectively and used in place of synthetic products.

Additionally, synthetic carotenoid produce is used for aquaculture to supplement animal feed; it accounts for a fifth of the feed expenditure (Quan and Turner, 2009). Astaxanthin is specifically included to increase cultured animal pigmentation. Hence, the colourant obtained from shrimp discard may be used for salmonoid and crustacean feed colouring (Mezzomo and Ferreira, 2016). Such opportunities have augmented demand for pigments and colouring agents derived from a natural material, attracting numerous researchers to assess the use of natural material for colouring products. Research has been conducted to assess and characterise the astaxanthin constitution pertaining to Penaeid shrimps (Alimba and Faggio, 2019; El-Bialy and Abd El-Khalek, 2020; Irna et al., 2017; Mohd Hatta and Othman, 2020).

Extraction Techniques for Astaxanthin Production

Numerous extraction techniques based on chemical, alkaline, ultrasound, oil, soxhlet, or supercritical fluids may be used for producing astaxanthin (Sun, Sangkatumvong, & Shung, 2006; Mezzomo et al., 2011; Sui et al., 2015). Every technique has some benefits and drawbacks. For example, carotenoid extraction is typically performed using chemical techniques because they are straightforward and cost-effective (Farre et al., 2015). Solvent-based extraction is usually based on the affinity characteristics of the solvent and sample. Carotenoids possess low polarity; hence, low-polarity solvents like hexane and petroleum ether provide for efficient extraction. However, several carotenoids comprise polar regions, thereby requiring careful solvent selection to facilitate better yield (Arvayo-Enríquez et al., 2013).

The use of extraction techniques is a significant aspect influencing product quality (Mezzomo et al., 2013). Conventional techniques require large quantities of acid and alkaline substances that impact astaxanthin yield and create environmental pollution (Sui et al., 2015). Moreover, the organic solvents used for extraction might be present in trace amounts in the processed products that may affect human health (Mezzomo & Ferreira, 2016). Additionally, standard production techniques typically require more solvent and time; simultaneously, substance degradation is also challenging (Herrero, Cifuentes, and Iban, 2006). Hence, green techniques for extracting carotenoids warrant additional research. Scholars have shown increased interest in this field, considering the number of natural products demanded by global markets.

Therefore, these challenges have been addressed by using the supercritical fluid extraction (SFE) technique. It provided better yield in a shorter duration and preserved compound quality (Santos et al., 2011; Silva, Gamarra, Oliveira & Cabral, 2008). Some studies have evaluated using vegetable oils to extract carotenoids from shrimp waste. Oils obtained from soybean, flaxseed, palm, and sunflower are used as co-solvents for supercritical carbon dioxide extraction (Sachindra & Mahendrakar, 2005; Pu, Bechtel, & Sathivel, 2010). SFE is a well-known technique with established benefits compared to conventional methods (Cadoni et al., 2000; Silva, Rocha-Santos, and Duarte, 2016; Mezzomo et al., 2013). Carbon dioxide is non-inflammable, non-toxic, and cost-effective; therefore, it is considered critical for several industrial processes.

Furthermore, industrial level carbon dioxide recycling can help regulate overall gas requirement, thereby facilitating cost-effective operation and reducing carbon footprint. Considering the non-polar molecular characteristics of CO_2, it is impractical to isolate polar substances; thereby, modifiers or co-solvents like isopropanol, ethanol, methanol, or hexane are required in varying concentrations. After introducing these substances to the supercritical extraction process at a specific flow, it is expected to enhance sample solubility and supercritical CO_2 polarity (Durante, Lenucci, and Mita, 2014). Considering solvent toxicity level, ethanol was selected as the co-solvent because it is effective and practical to use. It is categorised in the Generally Recognised as Safe (GRAS) class; hence, there is little human health or environmental concern for its applications in the food industry.

Carbon dioxide has critical pressure and temperature values of 72.9 atm and 31.3°C, respectively; hence, it is feasible to retain and preserve samples with high heat sensitivity (Durante, Lenucci, and Mita, 2014). It is possible to reduce extraction time using SFE because supercritical fluid possesses low viscosity but high diffusivity, facilitating fluid diffusion through the test specimens (Herrero, Cifuentes, and Iban, 2006).

Additionally, regulating temperature and pressure facilitates changing fluid density, thereby enhancing solubility. Also, the use of SFE for selective extraction is performed using an approach formulated through systematic experiment design. Sample variable response can be regulated by changing aspects like pressure, time, and temperature. It is feasible to control output level quality and obtain the required purity level. Hence, green techniques offer efficacious substitutes for use in several industries like pharmaceutical, food, dye, cosmetic, beverage, and others (Cadoni et al., 2000).

High-pressure processing (HPP) is another latest extraction technique considered safe for R&D and facilitates lesser destruction of bioactive substances when processed (Tadapaneni et al., 2014). The food industry is specifically interested in HPP because it can sustain food nutrients, preserve quality, sensory characteristics, and increase shelf life. These characteristics are because of the negligible influence of covalent bonds in compounds with less molecular mass. Moreover, thermo-labile substances are sensitive, and non-thermal techniques facilitate quality preservation (Ahmed and Ramaswamy, 2006). Specimen permeability is increased due to pressure-enhanced mass transfer (Xi, 2006).

Some studies assessed the HPP extraction technique. Xi (2013) evaluated HPP use for isolating active substances from plants. Sánchez et al., (2014) and McInerney et al., (2007) extracted plant-based carotenoids using HPP to determine overall carotenoid content. When HPP commences, there is a pressure differential between the external and internal cell membrane, causing gradual transfer until a concentration equilibrium is reached (Xi, 2006); this process is not time intensive. Like SFE, using HPP for extraction facilitates a noteworthy substance yield despite less time requirement.

Sample pre-processing is a vital step that affects extraction efficiency. Ovens are typically used for dehydrating samples; however, this process might cause thermolabile substances to have degraded or changed chemical characteristics, nutritional aspects, and colour (Durante, Lenucci, and Mita, 2014). Wang, Yang, Yan, & Yao (2012) indicated that total carotenoid concentration in raw pumpkin reduced by 65% after thermal treatment at 70 °C. Therefore, in order to prevent the degradation caused by heat treatment, freeze-drying is considered a better alternative because it preserves quality. The present study indicates similar outcomes. Powdered specimens subjected to freeze-drying had greater overall astaxanthin than oven-dried samples.

On the other hand, Perdigão, Vasconcelos, Cintra, & Ogawa (1995) indicated that heat-treated specimens lead to better extraction yield than cooking-free extraction. The researchers attribute this phenomenon to the thermal breaking of the protein-carotenoid complex, thereby increasing the extraction rate of the produced carotenoids. Moreover, they indicated that specimens having lesser moisture had better extraction yield, indicating the need for dehydrating the samples during pre-processing stages. Additionally, Mezzomo et al., (2011) suggested that powdered samples had significantly high carotenoid amount compared to non-powdered specimens. Powdered specimen particle size is less, leading to the higher exposed surface; hence, diffusion is enhanced because of higher mass transfer.

Antioxidant and Antimicrobial Properties of Astaxanthin

Carotenoids have excellent free-radical removing characteristics. They capture peroxyl radicals and quench singlet oxygen species (da Silva, Rocha-Santos, and Duarte, 2016). Carotenoid bioactive characteristics (like antioxidant properties) are typically affected by structural terminal groups, chemical arrangement, conjugated double bond presence, and oxygen-rich groups (Ligia et al., 2017). For example, shrimp waste-derived carotenoid astaxanthin is a lipophilic compound with potent antioxidant capabilities

(Sui et al., 2015). These characteristics are because of the distinct molecular arrangement that comprises keto moieties and hydroxyl on every ionone; moreover, these compounds have long chains of conjugated double bonds at the centre of the molecule (Delgado-vargas & Paredes-Lopez, 2000).

Soundarapandian, Shyamalendu, and Varadharajan (2014) researched activity assays to determine astaxanthin scavenging characteristics. The outcomes indicated 48% and 59% DPPH inhibition of the hard and soft crab shells, respectively, for the *C. lucifera* species. Research also suggested that crustacean-waste (by-product) derived astaxanthin bears potent antioxidant characteristics that potentially benefit health by offering protection against UV radiation and inflammation (Ushakumari & Ramanujan, 2012; Sindhu & Sherief, 2011).

Astaxanthin has superior antioxidant characteristics; its antioxidant potency is about ten times that of different carotenoids such as canthaxanthin, β-carotene, zeaxanthin, and lutein. Moreover, astaxanthin is 100 times more an antioxidant compared to α-tocopherol (Suganya & Asheeba, 2015). Superior antioxidant characteristics can be attributed to distinct chemical characteristics because of extended conjugated double bonds and the presence of ketone (C=O) and hydroxyl (OH) moieties on every ionone structure (Kishimoto et al., 2010). This distinct arrangement comprising a polar end ring provides astaxanthin with the ability to span the cell membrane bilayer. Explicitly, hydroxyl (OH) presence facilitates astaxanthin to capture reactive oxygen radicals from the surface of the membrane. At the same time, the polyene structure facilitates the reduction of oxidation chain processes inside the membrane.

Astaxanthin has a molecular arrangement comprising conjugated double bonds that attract electrons from reactive species, thereby facilitating free radical neutralisation (Rutz, Borges, Zambiazi, da Rosa, & da Silva, 2016). Hence, this superior antioxidant protects the entire cell and its components by inhibiting free-radical induced damage. The human body does not naturally synthesise carotenoids; hence, they must be fed externally using food (Suganya and Asheeba, 2015). Therefore, several academicians have attempted to evaluate natural substances possessing antioxidant characteristics that enable free radical capture, thereby slowing ageing, inflammation, and cancer-related processes (Licón et al., 2010).

Several studies have been conducted to assess the antimicrobial characteristics of astaxanthin. These studies evaluate how natural substances might help with disease prevention due to antimicrobial activity (Soureshjan and Heidari, 2014; Okmen et al., 2016). Such antimicrobial activity was established when research suggested potent effects against the typical disease-causing pathogens tested by the studies (Suganya & Asheeba, 2015). This medical property provides the preservative ability that hinders the development of foodborne pathogens responsible for food spoilage; moreover, beverage and food taste is enhanced due to such characteristics (Licón et al., 2010; Jinous et al., 2013). Food poisoning and diarrhoea are health conditions caused by *Staphylococcus aureus* and *Escherichia coli* (Jinous et al., 2013). These illnesses are characterised by food-caused conditions like meningitis, chronic renal disorders, watery or bloody stools, and immunologic, cardiovascular, or respiratory conditions (Mokhtarian, Mohsenzadeh & Khezri, 2004).

Astaxanthin Pigment Stability

Colour is a critical aspect of product aesthetics and contributes to marketability; moreover, colours themselves are accepted universally. Synthetic colouring agents are used extensively for numerous domains because of their economic properties and effectiveness. The use of synthetic colouring agents or additives is considered to adversely affect human health; colour toxicity might cause health issues (Samanta & Agarwal, 2009). Pigments or colouring agents are used for products in order to enhance visual appeal and characteristics, thereby enhancing product targeting. Hence, biological progress contributes to the use of natural colouring agents in place of synthetic compounds, thereby reducing the potential adverse toxicological effects on health and the environment.

Carotenoid characteristics concerning visual aspects and colour are associated with numerous environmental aspects like light, heat, pH, oxygen, and others. The conjugated double bond structure on the molecule defines colour by affecting the chromophore that absorbs light, substantiating the quantification and classification of carotenoid compounds (Rodriguez-Amaya, 2001). Sant'Anna et al., (2013) asserted that visible colour is reflected by carotenoids when there are at least seven conjugated double bonds (c.d.b) like 7 c.d.b in α-carotene that produce a soft yellow colour.

The colour degradation observed in the pigment indicates instability in highly acidic environments (Mohd Hatta and Othman, 2020). Acid is postulated to cause trans-cis isomerisation (Yip, Joe, Mustapha, Maskat, & Said, 2014). The stability of astaxanthin pH levels aligns with the research conducted by Rodriguez-Amaya (2001), carotenoids have better stability in alkaline environments than acidic environments. Moreover, acidic pH is understood to produce ion pairs leading to carotenoid carbocation (e.g., Car + AH ↔ (CarH$^+$···A$^-$) ↔ CarH$^+$ +A$^-$ (Boon et al., 2010).

Astaxanthin can be safely stored between 70 – 90 °C using different carriers; degradation is observed in the 120 - 150 °C temperature range (Mohd Hatta and Othman, 2020; Ambati et al., 2014). This observation might be attributed to esterified astaxanthin, which has better natural stability than free astaxanthin that is more vulnerable to oxygen, light, heat, and acidity. Such aspects contribute to degradation, isomerisation, oxidation, and other processes (Mezzomo et al., 2013; Mezzomo and Ferreira, 2016).

Carotenoid stability is impacted when exposed to light (Rodriguez-Amaya, 2001; Boon et al., 2010); trans-cis photoisomerization occurs on exposure (Yip et al., 2014; Rodriguez-Amaya, 2001). Absorption of light might cause the formation of single oxygen species causing carotenoid molecules to get excited, which might lead to chemically-caused degradation (Meléndez-Martínez et al., 2006; Boon et al., 2010). Therefore, it may be established that astaxanthin stability is strong influenced by light because of pigment photodegradation.

In the carotenoid context, production, consumption, storage, and delivery might inflict different environmental stresses on the compound, impacting its stability. Hence, there is a need for additional research to evaluate how such factors impact astaxanthin stability. Moreover, by dedicating additional efforts to researching methodologies to mitigate the challenges associated with natural pigments, there is a chance that the optimal conditions for pigment production, utilisation, and product development may be discovered. Additionally, it is crucial to evaluate the chemical processes between additives and natural colouring compounds since there is inadequate research on this subject. Therefore, additional research is recommended to discover stabilising agents that, when added, increase pigment stability.

CONCLUSION

Previous research works have studied the isolation of classification of lipophilic carotenoids (C_{40}), especially astaxanthin, using several natural sources. The evaluation of bioactivity (i.e., antibacterial, antifungal, and antioxidant) of such lipophilic astaxanthin isolates is established by the exceptional ability of astaxanthin derivates. These compounds have antioxidant and pro-vitamin A properties that are considered beneficial for health. Considering the rising health-conscious population, it is vital to consider the increasing consumption of astaxanthin products. Additionally, astaxanthin possesses colouring properties that might help prepare visually appealing food products. Hence, it is vital to evaluate astaxanthin colour stability consider several physical and environmental characteristics. Data concerning these aspects is expected to aid product formulation in numerous applied applications like beverages and food, polymers, nutraceuticals and pharmaceuticals. Hence, the focus should be on identifying additional natural sources of this colouring agent and a specific focus on identifying stabilising compounds that help enhance its capability to substitute synthetic compounds with naturally prepared compounds.

REFERENCES

Aberoumand, Ali. 2011. "A Review Article on Edible Pigments Properties and Sources as Natural Biocolorants in Foodstuff and Food Industry." *World Journal of Dairy & Food Sciences* 6 (1): 71–78.

Ahmed, Jasim, and Hosahalli S Ramaswamy. 2006. "Changes in Colour during High Pressure Processing of Fruits and Vegetables." *Stewart Postharvest Review* 5 (9): 1–8. https://doi.org/10.2212/spr.2006.5.9.

Alimba, Chibuisi Gideon, and Caterina Faggio. 2019. "Microplastics in the Marine Environment: Current Trends in Environmental Pollution and Mechanisms of Toxicological Profile." *Environmental Toxicology and Pharmacology* 68 (February): 61–74. https://doi.org/10.1016/j.etap.2019.03.001.

Ambati, Ranga Rao, Phang Siew Moi, Sarada Ravi, and Ravishankar Gokare Aswathanarayana. 2014. "Astaxanthin: Sources, Extraction, Stability, Biological Activities and Its Commercial Applications - A Review." *Marine Drugs* 12 (1): 128–52. https://doi.org/10.3390/md12010128.

Arvayo-Enríquez, Héctor, Iram Mondaca-Fernández, Pablo Gortáres-Moroyoqui, Jaime Lopez-Cervantes, and Roberto Rodríguez-Ramírez. 2013. "Carotenoids Extraction and Quantification: A Review." *Analytical Methods* 5 (12): 2916–24. https://doi.org/10.1039/c3ay26295b.

Bahukhandi, A., DHYANI, P., Bhatt, I. D., & Rawal, R. S. (2018). Variation in polyphenolics and antioxidant activity of traditional apple cultivars from West Himalaya, Uttarakhand. *Horticultural Plant Journal*, 4(4), 151-157.

Bohn, T. (2016). Bioactivity of carotenoids–chasms of knowledge. *Int J Vitam Nutr Res*, 10, 1-5.

Boon, Caitlin S, D Julian Mcclements, Jochen Weiss, and Eric A Decker. 2010. "Factors Influencing the Chemical Stability of Carotenoids in Foods Factors Influencing the Chemical Stability of Carotenoids in Foods." *Critical Reviews in Food Science and Nutrition* 50 (November 2016): 515–32. https://doi.org/10.1080/10408390802565889.

Britton, G., Liaaen Jensen, S., & Pfander, H. (2004). *Carotenoids Handbook*. Birkauser, Basel.

Cadoni, E, M R De Giorgi, E Medda, and G Poma. 2000. "Supercritical CO2 Extraction of Lycopene and Beta-Carotene from Ripe Tomatoes." *Dyes and Pigments* 44 (1): 27–32.

Chen, Xiaolin, Rong Chen, Zhanyong Guo, Cuiping Li, and Pengcheng Li. 2007. "The Preparation and Stability of the Inclusion Complex of Astaxanthin with β-Cyclodextrin." *Food Chemistry* 101 (4): 1580–84. https://doi.org/10.1016/j.foodchem.2006.04.020.

Delgado-vargas, Francisco, and Octavio Paredes-Lopez. 2000. "Natural Pigments : Carotenoids, Anthocyanins, and Betalains — Characteristics, Biosynthesis, Processing, and Stability Natural Pigments : Carotenoids, Anthocyanins, And." *Critical Reviews in Food Science and Nutrition* 40 (3): 173–289. https://doi.org/10.1080/10408690091189257.

Durante, Miriana, Marcello Salvatore Lenucci, and Giovanni Mita. 2014. "Supercritical Carbon Dioxide Extraction of Carotenoids from Pumpkin (Cucurbita Spp.): A Review." *International Journal of Molecular Sciences* 15 (1): 6725–40. https://doi.org/10.3390/ijms15046725.

Eggersdorfer, Manfred, and Adrian Wyss. "Carotenoids in human nutrition and health." *Archives of biochemistry and biophysics* 652 (2018): 18-26.

El-Bialy, Heba Abd Alla, and Hanan Hassan Abd El-Khalek. 2020. "A Comparative Study on Astaxanthin Recovery from Shrimp Wastes Using Lactic Fermentation and Green Solvents:An Applied Model on Minced Tilapia." *Journal of Radiation Research and Applied Sciences* 13 (1): 594–605. https://doi.org/10.1080/16878507.2020.1789388.

Farre, Gemma, Changfu Zhu, Gerhard Sandmann, Richard M Twyman, Teresa Capell, and Paul Christou. 2015. "Nutritionally Important Carotenoids as Consumer Products." *Phytochem Rev* 14: 727–43. https://doi.org/10.1007/s11101-014-9373-1.

Food and Agriculture Organization of the United Nations (FAO). (2014).*The State of World Fisheries and Aquaculture.*

Fortes, Cristina. 2006. "Carotenoids in Cancer Prevention." In *Carcinogenic and Anticarcinogenic Food Components*, edited by W. Baer-Dubowska, A. Bartoszek, and D. Malejka-Giganti, 283–302. United States of America: CRC Press.

Gómez-Estaca, J., M. M. Calvo, I. Álvarez-Acero, P. Montero, and M. C. Gómez-Guillén. 2017. "Characterization and Storage Stability of Astaxanthin Esters, Fatty Acid Profile and α-Tocopherol of Lipid Extract from Shrimp (*L. Vannamei*) Waste with Potential Applications as Food Ingredient." *Food Chemistry* 216: 37–44. https://doi.org/10.1016/j.foodchem.2016.08.016.

Grand View Research. (2017). *Carotenoids Market Analysis By Source (Natural, Synthetic), By Product (Beta-Carotene, Lutein, Lycopene, Astaxanthin, Zeaxanthin, Canthaxanthin), By Application (Food, Supplements, Feed, Pharmaceuticals, Cosmetics), And Segment Forecasts, 2018 – 2025,* Retrieved from https://www.grandviewresearch.com/industry-analysis/carotenoids-market

Gruber, K. (07 February 2013). *Nylons Made from Shrimps,* Retrieved from http://www.youris.com/Bioeconomy/Fisheries/Nylons_Made_From_Shrimps.kl.

Hashim, M., and S. Kathamuthu. 2005. *Shrimp Farming in Malaysia.* Tigbauan, Iloilo, Philippines: SEAFDEC Aquaculture Department.

Hermanns, A. S., Zhou, X., Xu, Q., Tadmor, Y., & Li, L. (2020). Carotenoid Pigment Accumulation in Horticultural Plants. *Horticultural Plant Journal*.

Herrero, Miguel, Alejandro Cifuentes, and Elena Iban. 2006. *Food Chemistry Sub- and Supercritical Fluid Extraction of Functional Ingredients from Different Natural Sources : Plants, Food-by-Products, Algae and Microalgae A Review* 98: 136–48. https://doi.org/10.1016/j.foodchem.2005.05.058.

Irna, C., I. Jaswir, R. Othman, and D. N. Jimat. 2017. "Antioxidant and Antimicrobial Activities of Astaxanthin from Penaeus Monodon in Comparison between Chemical Extraction and High Pressure Processing (HPP)." *International Food Research Journal* 24 (December): 508–13.

Jinous, Asgarpanaha, Darabi-Mahbouba Elahe, Mahboubib Arash, Mehrabb Rezvan, and Hakemivala Mojdeh. 2013. "In-Vitro Evaluation of Crocus Sativus L . Petals and Stamens as Natural Antibacterial Agents Against Food-Borne Bacterial Strains." *Iranian Journal of Pharmaceutical Sciences* 9 (4): 69–82.

Kishimoto, Y., Tani, M., Uto-Kondo, H., Iizuka, M., Saita, E., Sone, H., Kurata, H. & Kondo, K., (2010). Astaxanthin suppresses scavenger receptor expression and matrix metalloproteinase activity in macrophages. *Euro J Nutrit, 49* (2), 119–126.

Kittikaiwan, Prachanart, Sorawit Powthongsook, Prasert Pavasant, and Artiwan Shotipruk. 2007. "Encapsulation of Haematococcus Pluvialis Using Chitosan for Astaxanthin Stability Enhancement." *Carbohydrate Polymers* 70 (4): 378–85. https://doi.org/10.1016/j.carbpol.2007.04.021.

Li, Jian, Daling Zhu, Jianfeng Niu, Songdong Shen, and Guangce Wang. 2011. An Economic Assessment of Astaxanthin Production by Large Scale Cultivation of *Haematococcus Pluvialis*. *Biotechnology Advances*. Vol. 29. Elsevier Inc. https://doi.org/10.1016/j.biotechadv.2011.04.001.

Licón, Carmen, Manuel Carmona, Silvia Llorens, Maria Isabel Berruga, and Gonzalo L. Alonso. 2010. "Potential Healthy Effects of Saffron Spice (*Crocus Sativus L. Stigmas*) Consumption." *Functional Plant Science and Biotechnology* 4 (Special Issue 2): 64–73.

Ligia, A. C. C., Susan G. K., V. Francielo, Karen Y. F. K., Liliana I. C. Z., and C. C. Júlio. 2017. "Biotechnological Production of Carotenoids and Their Applications in Food and Pharmaceutical Products." In *Carotenoids*, 125–41. IntechOpen. https://doi.org/http://dx.doi.org/10.5772/57353.

Lin, Shen-fu, Ying-chen Chen, Ray-neng Chen, Ling-chun Chen, and Hsiu-o Ho. 2016. "Improving the Stability of Astaxanthin by Microencapsulation in Calcium Alginate Beads." *PLoS ONE* 11 (4): 1–10. https://doi.org/10.1371/journal.pone.0153685.

McWilliams, Andrew. 2018. *BCC Research Report Overview The: The Global Market Carotenoids.*

Meléndez-Martínez, Antonio J., George Britton, Isabel M. Vicario, and Francisco J. Heredia. 2006. "Relationship between the Colour and the Chemical Structure of Carotenoid Pigments." *Food Chemistry* 101 (3): 1145–50. https://doi.org/10.1016/j.foodchem.2006.03.015.

Meléndez-Martínez, A. J. (2019). An Overview of Carotenoids, Apocarotenoids, and Vitamin A in Agro-Food, Nutrition, Health, and Disease. *Molecular nutrition & food research*, *63*(15), 1801045.

Mezzomo, Natália, and Sandra R S Ferreira. 2016. "Carotenoids Functionality, Sources, and Processing by Supercritical Technology: A Review." *Journal of Chemistry* 2016: 1–17.

Mezzomo, Natália, Bianca Maestri, Renata Lazzaris Dos Santos, Marcelo Maraschin, and Sandra R. S. Ferreira. 2011. "Pink Shrimp (*P. Brasiliensis* and *P. Paulensis*) Residue: Influence of Extraction Method on Carotenoid Concentration." *Talanta* 85 (3): 1383–91. https://doi.org/10.1016/j.talanta.2011.06.018.

Mezzomo, Natália, Julian Martínez, Marcelo Maraschin, and Sandra R S Ferreira. 2013. "Pink Shrimp (*P. Brasiliensis* and *P. Paulensis*) Residue: Supercritical Fluid Extraction of Carotenoid Fraction." *Journal of Supercritical Fluids* 74: 22–33. https://doi.org/10.1016/j.supflu.2012.11.020.

Mohd Hatta, Farah Ayuni, and Rashidi Othman. 2020. "Carotenoids as Potential Biocolorants: A Case Study of Astaxanthin Recovered from Shrimp Waste." In *Carotenoids: Properties, Processing and Applications*, 289–325. Elsevier. https://doi.org/10.1016/b978-0-12-817067-0.00009-9.

Mokhtarian, H., Mohsenzadeh, M., Khezri, M. (2004). The survey on the bacterial contamination of traditional ice cream produced in Mashhad city. *Horizon Med Sci., 10*(1), 42-46.

Okmen, Gulten, Sukran Kardas, Duygu Bayrak, Ali Arslan, and Haldun Cakar. 2016. "The Antibacterial Activities of *Crocus Sativus* against Mastitis Pathogens and Its Antioxidant Activities." *World Journal Of Pharmacy and Pharmaceutical Sciences* 5 (3): 146–56.

Othman, Rashidi. 2009. *Biochemistry and Genetics of Carotenoid Composition in Potato Tubers.* Lincoln University.

Perdigão, N. B., Vasconcelos, F. C., Cintra, I. H. A., Ogawa, M. (1995). Extracão de carotenóides de carapacas de crustáceos em óleo, *Boletim Técnico Científico da CEPENE, 3* (1), 234. [Extraction of carotenoids from crustacean shells in oil, CEPENE *Technical Scientific Bulletin*]

Pu, Jianing, Peter J. Bechtel, and Subramaniam Sathivel. 2010. "Extraction of Shrimp Astaxanthin with Flaxseed Oil: Effects on Lipid Oxidation and Astaxanthin Degradation Rates." *Biosystems Engineering* 107 (4): 364–71. https://doi.org/10.1016/j.biosystemseng.2010.10.001.

Quan, Can, and Charlotta Turner. 2009. "Extraction of Astaxanthin from Shrimp Waste Using Pressurized Hot Ethanol." *Chromatographia* 70 (1–2): 247–51. https://doi.org/10.1365/s10337-009-1113-0.

Radzali, Shazana A, M Masturah, Badlishah S Baharin, O Rashidi, and Russly A Rahman. 2016. "Optimisation of Supercritical Fluid Extraction of Astaxanthin from Penaeus Monodon Waste Using Ethanol-Modified Carbon Dioxide." *Journal of Engineering Science and Technology* 11 (5): 722–36.

Rivera, S., F. Vilaró, and R. Canela. 2011. "Determination of Carotenoids by Liquid Chromatography/Mass Spectrometry: Effect of Several Dopants." *Analytical and Bioanalytical Chemistry* 400 (5): 1339–46. https://doi.org/10.1007/s00216-011-4825-6.

Rodrigo-Baños, Montserrat, Inés Garbayo, Carlos Vílchez, María José Bonete, and Rosa María Martínez-Espinosa. 2015. "Carotenoids from Haloarchaea and Their Potential in Biotechnology." *Marine Drugs* 13: 5508–32. https://doi.org/10.3390/md13095508.

Rodriguez-Amaya, Delia B. 2001. *A guide to carotenoid analysis in foods.* Washington, DC, USA: OMNI Research.

Rodriguez-Concepcion, M., Avalos, J., Bonet, M. L., Boronat, A., Gomez-Gomez, L., Hornero-Mendez, D., ... & Zhu, C. (2018). A global perspective on carotenoids: Metabolism, biotechnology, and benefits for nutrition and health. *Progress in lipid research, 70,* 62-93.

Rutz, J. K., Borges, C. D., Zambiazi, R. C., da Rosa, C. G., & da Silva, M. M. (2016). Elaboration of microparticles of carotenoids from natural and synthetic sources for applications in food. *Food Chemistry, 202*, 324–333.

Samanta, A. K., & Agarwal, P. (2009). Application of natural dyes on textiles. *Indian Journal of Fibre & Textile Research, 34*(December), 384–399.

Sachindra, N. M, and N. S Mahendrakar. 2010. "Stability of Carotenoids Recovered from Shrimp Waste and Their Use as Colorant in Fish Sausage." *Journal Food Sci Technol* 47 (February): 77–83.

Sant'Anna, Voltaire, Poliana Deyse Gurak, Ligia Damasceno Ferreira Marczak, and Isabel Cristina Tessaro. 2013. "Tracking Bioactive Compounds with Colour Changes in Foods - A Review." *Dyes and Pigments* 98 (3): 601–8. https://doi.org/10.1016/j.dyepig.2013.04.011.

Santos, Diego T, Carolina L C Albuquerque, and Maria Angela A Meireles. 2011. "Antioxidant Dye and Pigment Extraction Using a Homemade Pressurized Solvent Extraction System." *Procedia Food Science* 1 (0): 1581–88. https://doi.org/10.1016/j.profoo.2011.09.234.

Senthamil, L, and R Kumaresan. 2015. "Extraction and Identification of Astaxanthin from Shrimp Waste." *Indian Journal of Research in Pharmacy and Biotechnology* 3 (3): 192–95.

Silva, Rui P. F. F. da, Teresa A. P. Rocha-Santos, and Armando C. Duarte. 2016. "Supercritical Fluid Extraction of Bioactive Compounds." *Trends in Analytical Chemistry* 76: 40–51. https://doi.org/10.1016/j.trac.2015.11.013.

Sindhu, S, and P M Sherief. 2011. "Extraction, Characterization, Antioxidant and Anti-Inflammatory Properties of Carotenoids from the Shell Waste of Arabian Red Shrimp *Aristeus Alcocki*, Ramadan 1938." *The Open Conference Proceedings Journal* 2: 95–103.

Soundarapandian, P., Shyamalendu, R., & Varadharajan, D. (2014). Antioxidant activity in hard and soft shell crabs of *Charybdis lucifera*. *J Aquacul Res Dev, 5*(7), 1-5.

Soureshjan, Ehsan Heidari, and Mina Heidari. 2014. "In Vitro Variation in Antibacterial Activity Plant Extracts on Glaucium Elegans and Saffron (*Crocus Sativus* L) Onios." *Electronic Journal of Biology* 10 (3): 64–67.

Suganya, V, and S T Asheeba. 2015. "Antioxidant and Antimicrobial Activity of Astaxanthin Isolated From Three Varieties of Crabs." *International Journal of Recent Scientific* 6 (10): 6753–58.

Sui, Xiao, Rongyan Yue, Lan Wang, and Yuqian Han. 2015. "Process Optimization of Astaxanthin Extraction from Antarctic Kill (Euphausia Superba) by Subcritical R134a." In *3rd International Conference on Material, Mechanical and Manufacturing Engineering (IC3ME 2015)*, 47–53. Qingdao, China: Atlantis Press.

Sun, Lei, Suvimol Sangkatumvong, and K Kirk Shung. 2006. *"A High Resolution Digital Ultrasound System for Imaging of Zebrafish,"* 2202–5.

Sun, T., & Li, L. (2020). Toward the 'golden'era: the status in uncovering the regulatory control of carotenoid accumulation in plants. *Plant Science*, *290*, 110331.

Tadapaneni, Ravi Kiran, Hossein Daryaei, Kathiravan Krishnamurthy, Indika Edirisinghe, and Britt M Burton-freeman. 2014. "High-Pressure Processing of Berry and Other Fruit Products: Implications for Bioactive Compounds and Food Safety." *Journal of Agricultural and Food Chemistry* 62: 3877−3885.

Ushakumari, Uma Nath, and Ravi Ramanujan. 2012. "Astaxanthin from Shrimp Shell Waste." *International Journal of Pharmaceutical Chemistry Research* 1 (3): 1–6.

Wang, S. L., Chio, Y. H., Yen, Y. H. & Wang, C. L. (2007). Two novel surfactant-stable alkaline proteases from *Vibrio fluvalis* TKU005 and their applications. *Enzyme and Microbial Technology, 40*, 1213-1220.

Wang, L., Yang, B., Yan, & Yao, X. (2012). Supercritical fluid extraction of astaxanthin from *Haematococcus pluvialis* and its antioxidant potential in sunflower oil. *Innov. Food Sci. Emerg. Technol. 13*, 120–127.

Wurtzel, E. T. (2019). Changing form and function through carotenoids and synthetic biology. *Plant physiology*, *179*(3), 830-843.

Xi, Jun. 2006. "Effect of High Pressure Processing on the Extraction of Lycopene in Tomato Paste Waste." *Chemical Engineering & Technology* 29 (6): 736–39. https://doi.org/10.1002/ceat.200600024.

Yabuzaki, J. (2017). Carotenoids Database: structures, chemical fingerprints and distribution among organisms. *Database, 2017*.

Yan, N., and X. Chen. 2015. *Don't Waste Seafood Waste*. Macmillan Publishers, August 2015.

Yip, Wu Hon, Lim Seng Joe, Wan Aida Wan Mustapha, Mohamad Yusof Maskat, and Mamot Said. 2014. "Characterisation and Stability of Pigments Extracted from *Sargassum Binderi* Obtained from Semporna, Sabah." *Sains Malaysiana* 43 (9): 1345–54.

Zheng, X., Giuliano, G., & Al-Babili, S. (2020). Carotenoid biofortification in crop plants: citius, altius, fortius. *Biochimica et Biophysica Acta (BBA)-Molecular and Cell Biology of Lipids*, 158664.

In: What to Know about Carotenoids
Editor: Robert M. Albert
ISBN: 978-1-68507-105-9
© 2021 Nova Science Publishers, Inc.

Chapter 2

CHALLENGES IN ORAL BIOAVAILABILITY OF LUTEIN

Ishani Bhat, Saisree Iyer, Vanessa Fernandes, Divyashree M and Bangera Sheshappa Mamatha[*]
Nitte University Center for Science Education and Research, Paneer Campus, Deralakatte, Mangalore, Karnataka, India

ABSTRACT

Lutein is a xanthophyll carotenoid having a C40 isoprenoid (conjugated double bonds) backbone with oxygen-containing rings at both ends and is responsible for yellow-orange color in many fruits and vegetables. The conjugated double bonds confer lutein potent anti-oxidative property to manifest considerable biological activities that provide specific health benefits. Since lutein is synthesized only in photosynthetic organisms, it is categorized as an important dietary carotenoid to humans. On oral consumption of lutein, the effective concentration available at the target site depends on the amount consumed, absorbed, transported and metabolized within the host.

---

[*] Corresponding Author's E-mail: mamatha.bs@nitte.edu.in.

Further, the bioactivity of lutein is determined by the compound's bioaccessibility (the amount of lutein liberated from the food matrix during digestion) and bioavailability (the amount of lutein that enters the site of function). Since lutein is light and heat-sensitive pigment, processing conditions such as cooking, frying, and baking on dietary sources can reduce its levels before consumption. Several intrinsic and extrinsic factors influence the oral bioavailability of lutein. The key factors that influence carotenoid absorption are jointly abbreviated as SLAMENGI. It includes *S*pecies of the carotenoid, *L*inkage of the molecules, *A*mount of carotenoid consumed, *M*atrix surrounding the carotenoid, *E*ffectors of absorption and bioconversion, *N*utrition status of the host, *G*enetic variations of the host, and mathematical *I*nteractions. Lutein is a fat-soluble phytochemical that has been associated with the absorption of other lipophilic molecules like dietary lipids. It is released from the food matrix when ingested orally due to mechanical disruption and gastrointestinal juices. However, the oral bioavailability of lutein depends on its solubilization, degradation and rate of absorption. Additionally, variations in the genes that encode proteins involved in the absorption, transportation and metabolization of lutein within the host also modulate the oral bioavailability of lutein. This chapter is henceforth a compilation of the challenges encountered in the oral bioavailability of lutein.

Keywords: lutein, bioaccessibility, intestinal absorption, oral bioavailability, dietary challenges

1. INTRODUCTION

Carotenoids are tetra-terpenoids with a skeleton made up of 40 carbon atoms and 8 isoprene units that are separated into two categories: carotenes (pure unsaturated hydrocarbons) and xanthophyll carotenoids (oxygenated carotenoids). The human body has roughly 40 carotenoids out of the 600 that have been found. Carotenoids are present in human serum, with lutein, lycopene, β-cryptoxanthin, β-carotene, and zeaxanthin accounting for 20, 20, 10, 8, 6 and 3% of total serum carotenoids, respectively (Khachik et al., 1992; Khachik et al., 1997). Chemically, lutein is β,ε-carotene-3,3-diol ($C_{40}H_{56}O_2$) and lipophilic. Lutein has a strong yellow-orange color due to the presence of conjugated -C=C- bonds. These bonds enable free electron movement in lutein to absorb light from the blue region (450-485 nm) of the visible spectrum (Mora-Gutierrez et al., 2018; Yi et al., 2016). The isoprene backbone and double bonds in the lutein structure can exist in the *all-trans* or *cis* configurations (E/Z conformations) and lutein co-exists with its stereoisomer zeaxanthin. However, the *cis* isomer of lutein is thermodynamically less stable than the *all-trans* (E) configuration (Yang et al., 2018). Lutein consumption improves human eye health, prevents cardiovascular diseases and many types of cancer (Dwyer et al., 2001; Demmig-Adams and Adams 2002; Heber and Lu 2002; Landrum and Bone 2001). Food consumption has been correlated to age-related macular degeneration more than two decades ago (Goldberg et al., 1988). There has been extensive research on the importance of dietary nutrients in age-related macular degeneration ever since. Lutein is not synthesized in the human body. Thus, dietary intake is the only option to improve its concentrations in the macula. Lutein when consumed as supplements regularly improves the pigment density in the macula, which can reduce the occurrence of Age-related Macular Degeneration (Huang et al., 2015a; Huang et al., 2015b; Peng et al., 2016). Consumption of a lutein-rich diet can thereby minimize the risk of developing age-related macular degeneration (Eisenhauer et al., 2017). The amount of lutein consumed, its gastrointestinal absorption and metabolism by the host equally contribute to its active quantity of orally consumed lutein at the functional site. The abbreviation SLAMENGI rightly describes all the factors that affect lutein absorption (West and Castenmiller, 1998). The expansion refers to *s*pecies of the carotenoid, *l*inkage of the molecules, *a*mount of carotenoid consumed, *m*atrix surrounding the carotenoid, *e*ffectors of absorption and bioconversion, *n*utrition status of the host, *g*enetic variations of the host, and mathematical *i*nteractions as the factors. This chapter elaborately addresses the extrinsic factors affecting the bioaccessibility and bioavailability of lutein.

2. SOURCES OF LUTEIN AND ITS SIGNIFICANCE IN THE HUMAN BODY

Lutein is an oxygenated carotenoid that belongs to the non–provitamin A group called xanthophylls. It is found in abundance in green leafy vegetables and in yellow foods (Perry, Rasmussen, and Johnson, 2009). Lutein is found in plentiful sources including leafy vegetables, fruits, yellow foods, grains, cereals, flowers, and micro and macroalgae. Since lutein is synthesized only in photosynthetic organisms the daily requirement of lutein in humans should be provided through diet (Eisenhauer et al., 2017). However, lutein is also found in other non-photosynthetic sources like egg yolks and meat, which are considered as indirect sources.

2.1. Photosynthetic Sources

Xanthophyll carotenoids like lutein and zeaxanthin are commonly found in green leafy vegetables like kale, spinach, broccoli, parsley, peas and lettuce (Perry, Rasmussen and Johnson, 2009). Other vegetables and fruits like mango, pumpkin, corn, peaches, and other vegetables and nuts such as peas, asparagus, and pistachios are also identified as good sources of lutein (Table 1). Almost 70% of the total carotenoid content in corn is contributed by lutein and zeaxanthin (Kean, Hamaker and Ferruzzi, 2008). The lutein content (dry basis) of capsicum (green, red, and yellow peppers), yellow zucchini, kenaf, ivy gourd, broccoli, small red pumpkins, leeks, and carrots ranges from 0.033 to 0.419% (Aruna, Mamatha, and Baskaran, 2009; Mamatha, Sangeetha and Baskaran, 2011). Green leafy vegetables contain 0.012 to 0.015% of lutein. Among dietary grains, chickpea contains high levels of lutein (0.002 mg per 100 g of dry sample). Relatively high levels of lutein and zeaxanthin are found in einkorn, Khorasan and durum wheat and corn and their food products (Abdel-Aal et al., 2007; Abdel-Aal et al., 2010; de la Parra, Serna Saldivar and Liu, 2007;

Maiani et al., 2009). Few varieties of North American wheat bread like Catoctin and Pioneer also contain lutein and zeaxanthin (Humphries and Khachik, 20003). However, in comparison, other green-harvested wheat bread like Freekeh have substantially higher lutein and zeaxanthin.

Table 1. Sources of lutein and zeaxanthin

Sources	Lutein (µg/g)	Zeaxanthin (µg/g)	References
Vegetables			
Capsicum (green)	4193.7	-	(Mamatha, Sangeetha and Baskaran 2011)
Capsicum (red)	2752.5	-	
Zucchini (yellow)	2097.9	-	
Kenaf	1042.4	45.90	(Aruna, Mamatha, and Baskaran 2009)
Broccoli	2726	-	
Pumpkin	1062	2.78	
Spinach	59.3–79.0	-	(Maiani et al., 2009)
Corn	21.9	10.3	(Abdel-Aal et al., 2010)
Kale	48.0–114.7	-	(Maiani et al., 2009)
Parsley	64.0–106.5	-	(Maiani et al., 2009)
Cereals			
Flaxseed	1.8	0.005	(Mamatha, Sangeetha and Baskaran 2011)
Foxtail	1	0.002	
Pearl millet	0.9	0.004	
Finger millet	0.8	0.002	
Pulses and legumes			
Chickpea (split)	1.9	0.005	(Mamatha, Sangeetha and Baskaran 2011)
Red gram (split)	1.9	0.001	
Lentil (split)	10	0.004	
Cowpea	9.1	0.002	
Oils			
Mustard	7.7	1.20	(Aruna, Mamatha and Baskaran 2009)
Palm	0.1	-	
Macroalgae			
Bryopsis sp.	4.06	1.6	(Bhat, Haripriya, Jogi and Mamatha, 2021)
Ceramium sp.	3.26	0.66	
Chaetomorpha antennia	141.3	34.58	
Cladophora sp.	248.67	50.20	
Gracilaria corticata	0.26	0.65	
Grateloupia sp.	166.58	36.34	
Grateloupia filicina	18.38	2.16	
Sargassum whighti	0.46	-	
Ulva compressa	4.71	3.86	
Ulva fasciata	0.90	0.25	
Ulva lactuca	23.54	12.14	
Ulva prolifera	10.23	9.47	
Nuts			
Pistaschio	7.7–49.0		(Maiani et al., 2009)

2.2. Algal Sources

Both micro and macroalgae have been used for lutein extraction. Marine macroalgae including Chlorophyta, Rhodophyta, and Phaeophyceae were reported to have varying amounts of carotenoids (Bhat, Haripriya, Jogi and Mamatha, 2021). Lutein was reported in *Cladophora* sp. (248.67 µg/g), *Grateloupia* sp. (166.58 µg/g), *Chaetomorpha antennia* (141.30 µg/g), *Gracillaria edulis* (2.99 mg/g), *Gracillaria tikvahiae* (8.86 µg/g), *Grateloupia filicina* (18.38 µg/g), *Sargassum wightii* (0.46 µg/g), *Ulva compressa* (4.68 µg/g), *Ulva lactula* (23.54 µg/g) and *Ulva prolifera* (10.23 µg/g) (Bhat, Haripriya, Jogi and Mamatha, 2021; Othman et al., 2018; Rosemary et al., 2019). Few microalgae also have been used for the production of lutein like *Auxenochlorella prototothecoides* (34.13 mg/L), *Chlorella sorokiniana* (3.96 mg/L/day) and *Chlorella zofingiensis* (13.81 mg/g) (Chen, Ho, Liu and Chang, 2017; Huang et al., 2018; Xiao et al., 2018).

2.3. Non-Photosynthetic Sources

Eggs, fish, and crustaceans are also good sources of lutein but they are classified as indirect sources because the lutein in these foods is derived from plant sources. 17.17 µg/g of lutein and 12.52 µg/g zeaxanthin is present in chicken egg yolk (Handelman et al., 1999). However, the type of feed given to the chicken decides the lutein accumulation in the eggs (Nolasco et al., 2021). Although fruits and vegetables are a direct source of lutein, yolk from chicken eggs appear to contain more lutein (Mangels et al., 1993; Schaeffer, Tyczkowski, Parkhurst and Hamilton, 1988). This is merely due to improved bioavailability (section 3.2.2, 4 and 5.1) from the high-fat content in eggs.

2.4. Microbial Sources

Generally, carotenoids are biosynthesized by all autotrophic marine organisms like bacteria and archaea, algae and fungi. Non-photosynthetic marine organisms are unable to synthesize carotenoids *de novo*. Common bacterial species reported for carotenoid production include *Agrobacterium* spp., *Arthrobacter* spp., *Chromobacterium* spp., *Flavobacterium* spp., *Micrococcus* spp., *Pseudomonas aeruginosa*, *Rheinheimera* spp., and *Serratia marcescens* which are important for industrial carotenoid production (Ram et al., 2020; Galasso, Corinaldesi and Sansone, 2017). Competitive industrial demand for carotenoids production uses bacteria which forms another alternative for plant-based carotenoids, due to the short life cycle and ease of maintenance during fermentation. Additionally, genetically engineered organisms are used to improve the production of carotenoids from low-cost substrates or wastes into valuable carotenoids (Saini and Keum, 2019). Carotenoids are associated with proteins in the gram-negative bacterial membrane hence enzymatic extraction of carotenoids improves extraction proficiency, is environmentally friendly and reduces the solvent usages for extraction procedures (Ram et al., 2020).

2.5. Significance in the Human Body

Dietary carotenoids, lutein and zeaxanthin, are accumulated in the human retina (Krinsky, Landrum and Bone, 2003) to absorb blue light, enhance visual acuity and forage harmful reactive oxygen. Lutein and zeaxanthin, along with their common metabolite *meso*-zeaxanthin, are referred to as macular pigments (Bone, Landrum, Fernandez and Tarsis, 1988). The ratio between lutein, zeaxanthin and *meso*-zeaxanthin changes as the eccentricity moves away from fovea (Bone, Landrum, Fernandez and Tarsis, 1988; Handelman, Dratz, Reay and Van Kuijk, 1988; Landrum and Bone 2001). Although lutein and zeaxanthin were also detected in prenatal eyes, however, they did not form a visible yellow spot. Lutein plays an important role in protecting the retina from oxidative damage that is caused by exposure to blue light. Lutein quenches singlet oxygen molecules that are by-products of inflammation by forming zeaxanthin radicals, thus decreasing inflammation (Dall'Osto et al., 2017). The ratio of lutein to zeaxanthin differs between infants and adults. However, there is no change in the concentration of lutein and zeaxanthin with progressing age (Bone, Landrum, Fernandez and Tarsis, 1988). In infants, lutein predominates over zeaxanthin in the fovea, and the opposite is true after 3 years of age (Bone, Landrum, Fernandez and Tarsis, 1988; Moukarzel, Bejjani and Fares, 2009). It is observed that lutein plays a crucial role in prenatal human development (Panova et al., 2017). In a placebo-controlled clinical trial, patients with age-related macular degeneration for 2 years showed a noteworthy increase in their serum lutein levels and macular pigment optical density in those patients supplemented with lutein and zeaxanthin (Huang et al., 2015a). Lutein induces the expression of antioxidant enzymes like superoxide dismutase and promotes tight-junction repair in retinal pigment epithelial cells in experimental mice models (Kamoshita et al., 2016). Retinal sensitivity and visual performance were also enhanced by supplementation of lutein in patients with early age-related macular degeneration (Huang et al., 2015b). In the same study, no detectable adverse effects were seen when the subjects were dosed with 10 mg/day for long-term treatment. Among other carotenoids, lutein plays an important role in neural development and is most abundantly found in infant brains (Vishwanathan, Kuchan, Sen and Johnson, 2014). Reduced levels of macular pigments were associated with a greater occurrence of age-related macular degeneration in patients with Alzheimer's disease (Nolan et al., 2014). This indicates a positive effect of lutein on cognitive functions. This is due to its protective action against the oxidation of DHA, which is an important omega-3 fatty acid found in all vital parts of the brain. Lutein supplementation has shown to improve memory, attentiveness and reasoning ability in young adults (Renzi-Hammond et al., 2017). Other studies report that lutein consumption helps lower serum cholesterol, hepatic cholesterol, triglyceride levels (Murillo et al., 2016, Qiu et al., 2015) and susceptibility of low-density lipoproteins to lipid peroxidation (Kishimoto et al., 2017). It was also seen that lutein can reduce testicular damage and oxidative stress in diabetic rats (Fatani et al., 2015). Lutein repressed antivascular endothelial growth factor and further showed angioprotective efficacy in diabetic retinopathy in a study in rats (Sharavana and Baskaran, 2017). Lutein acts as a chemoprotective agent against atherosclerosis in mice by showing improved lipid metabolism (Han et al., 2015).

3. EXTRINSIC CHALLENGES AFFECTING BIO-ACCESSIBILITY OF LUTEIN

Orally consumed lutein undergoes series of mechanical and chemical digestive processes that release the lutein from the food matrix, making way for its absorption. However, the processing conditions underwent by the source before consumption also contribute to modifying the bioaccessibility of lutein (Figure 1).

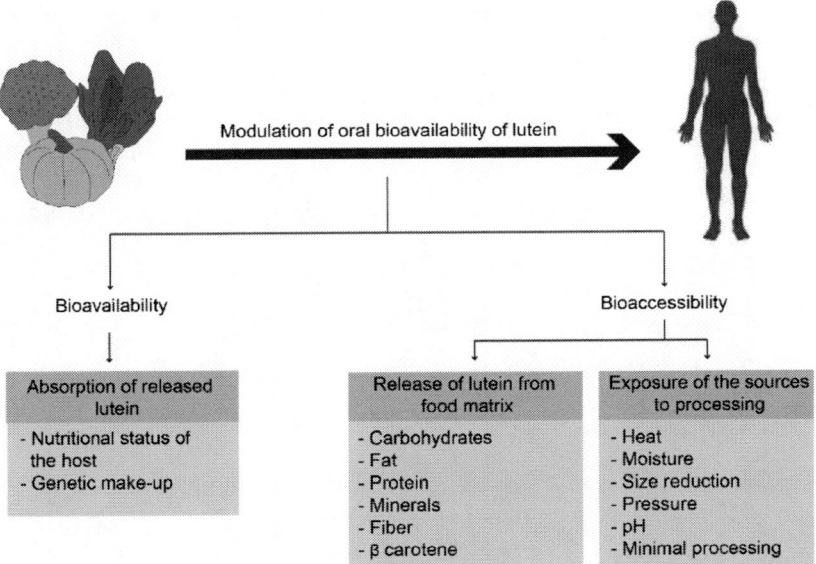

Figure 1. Challenges posed by extrinsic and intrinsic factors on oral bioavailability of lutein.

Table 2. Effect of processing conditions on lutein concentration

Processing conditions	Specifications	Source	Effect on lutein	Reference
Boiling	Short time (4-8 min)	Spinach	-	(Chung, Leanderson, Gustafsson and Jonasson, 2019)
	Medium time (16-30 min)		- -	
	Long time (up to 90 min)		- - -	
	5 min	Broccoli	+++	(Hwang and Kim, 2013)
	10 min		++++	
	NA	Mugwort	+	(Komuro et al., 2017)
Wet cooking	NA	Sorghum flour	+++	(de Morais Cardoso et al., 2014)
Microwave heating	10 min	Broccoli	++++	(Hwang and Kim, 2013)
Microwave re-heating	After boiling	Spinach	++	(Chung, Leanderson, Gustafsson and Jonasson, 2019)
	After steaming		++	
Steaming	NA	Spinach (Whole leaves)	-	(Eriksen, Luu, Dragsted and Arrigoni, 2016)
	NA	Minced spinach	+	
	Short time (4-8 min)	Spinach	- -	(Chung, Leanderson, Gustafsson and Jonasson, 2019)
	Medium time (16-30 min)		- -	
	10 min	Broccoli	+++	(Hwang and Kim, 2013)
Blanching	5 min at 70° C	Corn	+	(Mamatha, Arunkumar and Baskaran 2012)
		Onion stalk	+	
		Broccoli	×	
		Capsicum	+	
Stir frying	NA	Whole spinach leaves	-	(Eriksen, Luu, Dragsted and Arrigoni, 2016)
	NA	Minced spinach	++	
Pan frying	Short time (4-8 min)	Spinach	- -	(Chung, Leanderson, Gustafsson and Jonasson, 2019)
	Medium time (16-30 min)		- -	

Processing conditions	Specifications	Source	Effect on lutein	Reference
Roasting	Short time	Pistachios	+	(Pumilia et al., 2014)
	Long time		+	
Dehydration/Drying	NA	Mugwort	+	(Komuro et al., 2017)
	After milling, heat and moisture treatment	Orange maize flour	-	(Beta and Hwang, 2018)
		Corn	+	(Mamatha, Arunkumar and Baskaran 2012)
		Onion stalk	-	
		Broccoli	++	
		Capsicum	-	
Dry cooking	NA	Sorghum flour	+	(de Morais Cardoso et al., 2014)
Liquefaction	Blended with distilled water	Spinach	++	(Chung, Leanderson, Gustafsson and Jonasson, 2019)
	Blended with distilled water and heavy cream		++++	
Cutting	After boiling	Mugwort	+	(Komuro et al., 2017)
Peeling	Hand peeling	Carrot	--	(Ma et al., 2015)
	Hot water peeling		-	
	Lye peeling		-	
	Composite phosphate peeling		-	
High pressure processing		Carrot juice	-	(Stinco et al., 2019)
Milling		Corn	+	(Mamatha, Arunkumar and Baskaran, 2012)
		Onion stalk	--	
		Broccoli	+	
		Capsicum	×	

Note: Increase up to 25%, 50%, 75% and >75% is denoted by +, ++, +++ and ++++ respectively. Decrease up to 25%, 50% and 75% is denoted by -, -- and --- respectively. No significant increase or decrease is denoted by ×.

3.1. Challenges Faced by Lutein during the Processing Conditions

Food has to be processed before consumption to improve its palatability, which involves processes that include heat, high pressure and size reduction. Additionally, pH and moisture content in the sources affects the stability of lutein (Table 2).

3.1.1. Harvesting Conditions

The major factor affecting the amount of lutein absorbed from food is the amount of lutein consumed through diet. However, the changes in the food source matrix and existing molecular linkages are highly influenced by the processing conditions that the dietary sources undergo. The decrease in the concentration of lutein in plant sources begins from the period of harvest. The highest concentration of lutein has been seen in green-leafy and yellow vegetables at maturity followed by a drop in concentration at hyper maturity (Lefsrud, Kopsell, Wenzel & Sheehan 2007; Ma et al., 2015).

3.1.2. Moist Heat Treatment

Major sources of lutein in our diet (eggs, spinach, kale, corn, broccoli, etc.) are commonly consumed after thermal processes like boiling, steaming, pan-frying, etc. Bioactive compounds in the vegetables, that are heat labile, are prone to denaturation and degradation during heat processing. The extent of its degradation, however, depends on the duration of processing. For example, reduction in the bioaccessibility of lutein from spinach on boiling has been reported to be time-dependent (Chung, Leanderson, Gustafsson and Jonasson, 2019). Boiling for longer durations (90 min; stew making) significantly reduced lutein liberation in comparison to short-term (4-8 min) or medium-term (16-30 min; soup making) heating. Remarkably, microwave re-heating was able to partially compensate for the loss. Besides, steaming whole spinach leaves caused reduced liberation of lutein (Eriksen, Luu, Dragsted and Arrigoni, 2016). A similar increase in the lutein content was reported when broccoli was steamed and microwaved for a short duration (Eriksen, Luu, Dragsted and Arrigoni, 2016). Boiling and dehydration usually lead to the loss of water-soluble proteins that concentrate lutein. Conversely, boiling and dehydrating mugwort did not reduce the lutein content indicating that lutein liberation is also dependent on the decomposition of its source matrix (Komuro et al., 2017). Short period blanching (70° C) increases the release of lutein from the matrix in corn and other vegetables (onion stalk, broccoli and capsicum) by a significant amount (Ranganathan, Mamatha and Baskaran, 2014; Mamatha, Arunkumar and Baskaran, 2012). Agitation and heat break the cellular matrix of the plant and increase the surface area facilitating the release of lutein (Boon, McClements, Weiss and Decker, 2010). But increased pressure and reduced cooking time can eventually lead to reduced losses in vegetables (Sanchez, Baranda and de Maranon, 2014).

3.1.3. Dry Heat Treatment

Lutein is a heat and light-sensitive pigment and when exposed to high-temperature conditions like cooking, baking and frying, a drop in its concentration can occur (Gutierrez-Uribe, Rojas-Garcia, Garcia-Lara and Serna-Saldivar, 2014; Shen et al., 2015). Dry heat treatment improves the bioaccessibility of lutein. Roasting pistachios for a short duration significantly increase the extractable concentration of lutein (Pumilia et al., 2014). However, roasting at higher temperatures for longer periods can do the reverse. Besides, stir-frying can double the lutein content in minced spinach (Eriksen, Luu, Dragsted and Arrigoni, 2016). However, dry heat treatment can be less detrimental to lutein degradation in comparison to moist heat treatment. Dry heating of ground raw maize reduced lutein degradation by 50% in comparison to prolonged moist heat treatment (Beta and Hwang, 2018). There is a large margin of difference between the retention rates of lutein during wet and dry cooking of sorghum flours (Beta and Hwang, 2018). The negative effect of moisture on lutein degradation during heat processing can be reduced by alternative dry heat treatment. The lutein content in maize is enhanced on nixtamalization however on conversion into tortillas by baking, the quantity of lutein is again reduced to its initial level (Gutierrez-Uribe, Rojas-Garcia, Garcia-Lara and Serna-Saldivar, 2014). Paling of foxtail millets, which are rich in lutein and zeaxanthin, is observed on cooking thus indicating a loss of yellow pigments (Shen et al., 2015).

3.1.4. Moisture and Dehydration

Removal of moisture and dehydration reduces lutein content due to residual peroxidase enzyme activity in vegetables (Mamatha, Arunkumar and Baskaran, 2012). Although moist heat treatment is found to reduce levels of lutein, short-term treatment before dehydration was reported to be beneficial. Blanching onion stalk and capsicum before drying improved their lutein levels (Mamatha, Arunkumar and Baskaran 2012). However, the opposite was reported in dried corn and broccoli.

3.1.5. Pre-Processing Techniques

Pre-processing techniques like washing, peeling, cutting, slicing and chopping minimally affect lutein levels. Peeling of carrots has been found to cause loss of lutein (Ma et al., 2015). While lutein losses incurred by hand peeling were attributed to irregular pulp loss due to manual error, losses incurred by lye peeling were related to its corrosive effects on lutein. Processing operations like milling, homogenizing, grinding, chopping, etc. applied for size reduction can significantly enhance the extractability of lutein from vegetables in comparison to drying processes. The extractability and retention levels of lutein can be further preserved by up to 20-115% when blanching is coupled with size reduction processes in vegetables like onion stalk and capsicum (Mamatha, Arunkumar and Baskaran 2012). Mincing of spinach before heat treatment also improved its lutein content (Eriksen, Luu, Dragsted and Arrigoni, 2016). However, the retention or loss of lutein during size reduction processes depends on the nature of the food source.

3.1.6. Application of Pressure

Other non-thermal processes like high-pressure processing, pureeing, juicing, etc. that produce liquefied products show a higher yield of lutein that is optimal for better liberation from the food matrix (Chung, Leanderson, Gustfsson and Jonasson, 2019). When lutein is liberated from the plant matrix, it is prone to degradation due to exposure to enzymes, oxidative species and other agents. Also, high-pressure processing can directly break down lutein (Stinco et al., 2019).

3.1.7. pH Changes

The pH range in food products and beverages is very versatile. While fruit juices and soft drinks are acidic, milk and milk substitutes are neutral beverages. Lutein in emulsions is completely unstable at pH 4-5, which causes droplet aggregation and creaming (Boon, MsClements, Weiss and Decker, 2010). Although lutein in emulsion at room temperature is slightly unstable at higher pH ranges, lutein shows high stability at pH 8. In acidic conditions, xanthophyll carotenoids undergo protonation of C atoms of the conjugated system resulting in accelerated degradation and isomerization of the compounds (Davidov-Pardo, Gumus and McClements, 2016). Due to the presence of hydroxyl groups in lutein that alter the protonation of C atoms, lutein is usually stable against changing pH.

3.2. Challenges Faced by Lutein in the Digestive Tract

To achieve the potential health benefits of lutein consumed from food, the following steps are essential – (1) lutein release from food matrix and transfer to micelles during digestion called 'bioaccessibility'; (2) uptake by enterocytes; (3) transport and packing into the chylomicrons for secretion into the lymph. Micellization efficiency is defined as the percentage of carotenoids transferred to the mixed micelles from the crude digesta, and is often used as a measure of bioaccessibility. Several factors affect the bioaccessibility of lutein like the interactions of lutein with the components of the food matrix-like carbohydrates, fats, proteins, fibers and other carotenoids (Goñi, Serrano and Saura-Calixto, 2006). However, the extent of lutein micellization from different foods depend upon the source matrix (Margier et al., 2018). The interaction of digestive enzymes with the food matrix during gastrointestinal digestion decides the fate of lutein, which is to be further absorbed. During gastrointestinal digestion, the food matrix is mechanically and chemically degraded to facilitate the transfer of lutein into mixed micelles.

3.2.1. Carbohydrates

Carbohydrates form a major portion of the food matrix of our diet and potatoes are one among the carbohydrate-rich foods. Studies report that there is up to 80% retention of lutein in yellow-fleshed potatoes after boiling (Burgos et al., 2013). Upon digestion study, it was found that there is a slight decrease in the lutein concentration after digestion from the gastric phase and a significantly higher decrease after digestion from the duodenal phase, thus suggesting poor stability in the duodenal environment. Due to this, very low concentration of lutein is retained for micellization. Thus, carbohydrates in the food matrix negatively affect the bioaccessibility of lutein.

3.2.2. Fat

The presence of fat plays a very important role in the bioaccessibility and absorption of lutein owing to its fat-soluble nature. Lutein, when added to a full fat spread, shows excellent plasma response on consumption (van het Hof, West, Weststrate and Hautvast, 2000). Excellent solubilization of lutein esters in the fat phase helps the release of intestinal esterases and lipases for micelle formation. Besides, when spinach was orally administered to rats with fats (butter, fish oil and olive oil), plasma lutein level was increased (Gleize et al., 2012). The increase was reported to be highest in butter in comparison to the other two oils. Hence, the bioaccessibility of lutein is negatively related to the degree of unsaturation of triacylglycerides (TAG) fatty acids. Additionally, the mean length of the fatty acid present in the TAG is negatively related to the bioaccessibility of lutein. However, micellar size increases with the addition of long-chain fatty acids to bile salts (Carey and Small, 1970). As micellar size and absorption of lutein are inversely related, long-chain fatty acids reduce the oral bioavailability of lutein.

3.2.3. Protein

Proteins are surface-active emulsifiers that facilitate emulsion stabilization by forming a physical barrier and absorbing the oil droplets to prevent coalescence during digestion. They can also influence lutein uptake by modulating the lutein micellization in fat droplets. The presence of low concentrations of whey protein isolate, soy protein isolate and sodium caseinate decrease the bioaccessibility of lutein to about 60%, 41% and 70% respectively (Iddir et al., 2020). The presence of gelatin also harms the bioaccessibility of lutein.

3.2.4. Mineral

Various minerals like sodium, calcium, magnesium, zinc etc. have varied effects on the bioaccessibility of lutein. While sodium improves micellization of lutein, calcium and magnesium have a significant negative effect on the same (Corte-Real et al. 2016; Corte-real et al. 2017). The binding of these minerals with fatty acids and bile salts could hinder the micelle formation and in turn affect the delivery of lutein to the enterocytes. Zinc shows no significant change in the micellization of lutein when its concentrations are within the physiological levels.

3.2.5. Fiber

Dietary fibers reduce the bioavailability of macronutrients which affect the absorption of lutein. They enhance the viscosity of gastric fluids to restrict the peristaltic mixing process, which limits the transport of enzymes to their substrates. The fiber in maize is located in the hull (51%) and germ (16%) fractions. Bioaccessibility of lutein in maize improves on milling due to the removal of fiber (Zhang et al., 2020). Fibers in vegetables also reduce the bioaccessibility of lutein from plant foods. They interfere with the micelle formation by partitioning bile salts and fats in their gel phase (van het Hof, West, Weststrate and Hautvast, 2000). Citrus pectin, wheat bran and oat bran also inhibit micelle formation thus reducing lutein bioaccessibility (O'Connell et al., 2008).

3.2.6. Carotenoids

Interactions of lutein with other carotenoids at the intestinal level may competitively reduce the absorption of lutein. Competition may occur at the level of micelle formation, intestinal uptake or lymphatic transport. Besides, simultaneous ingestion of various carotenoids may induce an antioxidant-sparing effect in the intestinal tract and result in increased levels of uptake of the protected carotenoids (van het Hof, West, Weststrate and Hautvast, 2000). Competitive effect between lutein and β-carotene is a well-known phenomenon (Zhang et al., 2020). Although lutein has been found to micellize better than β-carotene (O'Connell et al., 2008; Zhang et al., 2020), the reverse is true for intestinal absorption (Mamatha and Baskaran, 2011). However, the bioaccessibility of lutein does not wholly depend on the presence of other carotenoids.

4. ABSORPTION OF LUTEIN

Hydrophobic phytochemicals are absorbed into the systemic circulation similar to the dietary lipids (Figure 2). Lutein present in the food is released from the food matrix due to the impact of mechanical disruption and the action of gastrointestinal juices. The released lutein gets dispersed in lipids, forming emulsions, as it is unstable in aqueous phases. In due course, they turn into micelles composed of fatty acids, oleic acid, mono glycerol, cholesterol and phosphatidylcholine (Baskaran, Sugawara & Nagao, 2003). The lutein embedded micelles get absorbed into the intestinal mucosal cells of the duodenum through transmembrane proteins that are isoforms of scavenger receptor class B type 1 (SR-B1) and a cluster of differentiation 36 (CD36) (Borel et al., 2013; During, Doraiswamy & Harrison, 2008; Reboul et al., 2005). In recent times, a transmembrane protein NPC1L1 (Niemann-Pick C1-Like 1) was found to be associated with absorption of lutein (Sato et al., 2012) and possibly a few more unidentified transporters. In contrast to β-carotene which gets cleaved by β-carotene 15,15'- dioxygenase (BCO1) or β-carotene 9',10'-monooxygenase (BCO2) enzymes into retinol, lutein is not metabolized within the intestinal mucosal cell (Amengual et al., 2013; dela Sena et al., 2013). These enzyme phenotypes need to be studied for better clarity as competitive absorption is seen between the two carotenoids. Thereafter, lutein is incorporated in the chylomicrons present in the Golgi apparatus and transported to the liver via portal circulation (Borel et al., 1996). While the mechanism is not known, it is assumed that a carrier is required for hydrophobic lutein to cross the aqueous intracellular compartment. However, lutein can be incorporated directly into high-density lipoproteins (HDLs) through the apolipoprotein A1 dependent route (APOE1) via ATP binding cassette A1 (ABCA1) transporter (Niesor et al., 2014).

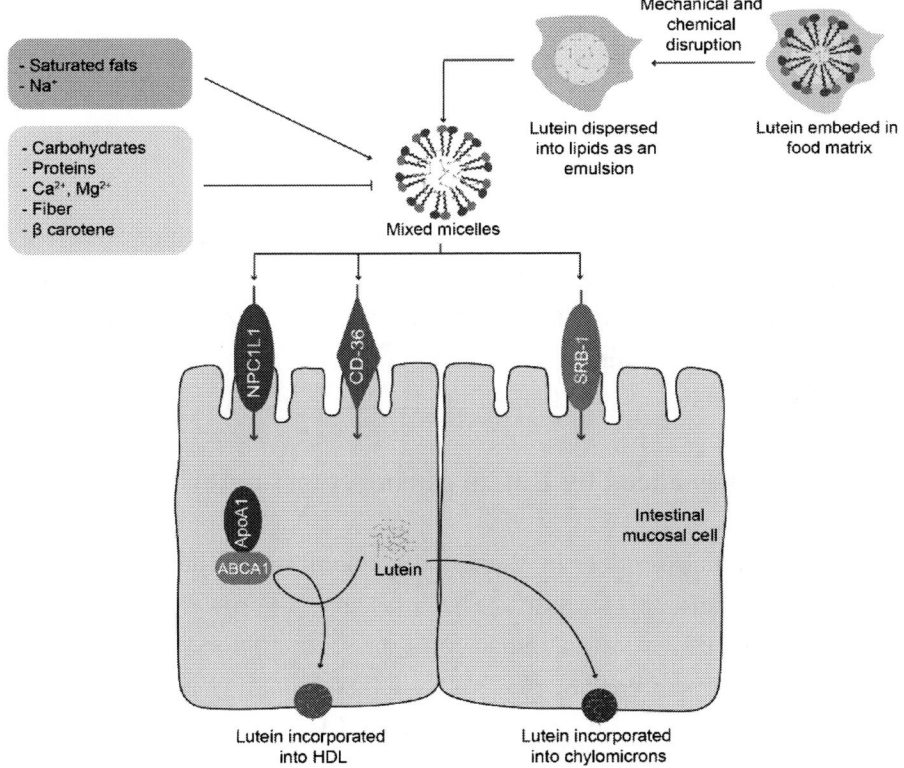

Figure 2. Gastrointestinal digestion and absorption of lutein as affected by the extrinsic and intrinsic factors.

For targeted transport to tissues like the eye and brain, lutein is loaded into HDL and carried via the vascular system (Thomas & Harrison, 2016). Amongst all the lipoproteins found in humans, lutein is typically dispersed highest in HDL (53%), then in low-density lipoproteins (LDL-31%) and least in very-low-density lipoproteins (VLDL-16%) (Parker, 1996). Each type of lipoprotein can carry a defined number of lutein molecules. 20 HDL molecules are needed to carry a single molecule of lutein whereas, 1 molecule of LDL is enough to carry a single molecule of lutein (Harrison, 2019). In humans, lutein is known to be strongly associated with HDL however, the mechanism is still not clear.

5. Intrinsic Challenges Affecting Bioavailability of Lutein

The absorption of lutein includes several steps, as described above. The overall bioavailability of lutein is affected by any factors that may interfere with any of these steps. The release of lutein and incorporation into mixed micelles may affect the absorption of lutein and thereby modulate its bioavailability.

5.1. Challenges Faced by Lutein during Intestinal Absorption

Lutein, on absorption in the gut, is taken up in the chylomicron micelles for further transport. The presence of dietary macro and micronutrients can have an impact on lutein absorption during nutrient utilization (Korobelnik et al., 2017). Lutein bioavailability is lowered when dietary fibers are added to lutein supplements (Riedl, Linseisen, Hoffmann and Wolfram, 1999; Mamatha and Baskaran, 2011). Moreover, competitive absorption between carotenoids can be a possible barrier to lutein uptake (Kostic, White and Olson, 1995; Reboul et al., 2007). In the presence of β-carotene, the absorption of lutein in the intestine is impeded

(Mamatha and Baskaran, 2011). The absorption of lutein is increased when it is combined with lipids, oils and lipoproteins. (Kijlstra, van der Made, Plat and Berendschot, 2017; Mamatha and Baskaran, 2011; Marriage et al., 2017; Nidhi, Mamatha and Baskaran, 2014; Roodenburg, 2000). The bioavailability of lutein is improved in foods that are processed to have a higher content of fat (Read, Wright and Abdel-Aal, 2015; Xavier, Carvajal-Lerida, Garrido-Fernandez and Perez-Galvez, 2018). The bioavailability of lutein is reduced when fat is removed from the dietary products (Xavier, Marcadante, Garrido-Fernandez and Perez-Galvez, 2014). Also, when lutein is combined with polyunsaturated fats in the diet, its bioavailability is reduced (Wolf-Schnurrbusch et al., 2015).

5.2. Host Factors That Modulate Oral Bioavailability of Lutein

In human plasma, the ratio of lutein/zeaxanthin is found to be 2:1 to 4:1. Its intestinal absorption pathway includes sequential protein involvements. The efficacy of absorption of lutein in the intestine is altered by variations in genes that code for these specific proteins. There are limited studies that have explored the association between lutein bioavailability and the genetic variations in genes encoding these proteins (Borel et al., 2011; Borel et al., 2014; Huebbe, Lange, Lietz & Rimbach, 2016; Yonova-Doing et al., 2013). As per the studies on inter-individual variability in some genes related to lutein absorption and transport, there are a few SNPs modulating lutein bioavailability in humans. Among the genes that code for proteins responsible for digestion, transport and metabolism of sterols, variations in the genes BCMO1 (also known as BCO1) and CD36 were found to be responsible to modulate lutein bioavailability (Borel et al., 2011). Subjects that carried allele TT rather than CT at the BCMO1 rs7501331 locus and allele CC rather than T at the CD36 rs13230419 locus showed a significantly low concentration of plasma lutein. However, significantly higher MPOD was observed in the subjects that carried allele TT at the BCMO1 rs7501331 locus and allele GG at the CD36 rs1761667 locus. Association between the lutein concentration in the serum and SNPs at the loci BCMO1 rs654851 and SCARB1 rs11057841 has been reported (Yonova-Doing et al., 2013). It was also noted that MPOD is associated with the SNPs at loci SCARB1 (scavenger receptor class B member 1) rs11057841, RPE65 (retinol isomerohydrolase) rs4926339, ABCA1 rs1929841 and FADS1 (fatty acid desaturase 1) rs174534. There is direct involvement of BCMO1, SCARB1 and ABCA1 in the intestinal absorption of lutein, while RPE65 is involved in supplying 11 -cis retinal chromophore, a pigment present in the human retina. Lutein is transformed by the isomerase coded by RPE65 (Shyam et al., 2017) into meso-zeaxanthin which is an important constituent of the macula, thus explaining its relation with MPOD. FADS1 is not directly involved in the absorption or transport of lutein. However, its modulatory action on the bioavailability of lutein can be explained as it is involved in converting α-linolenic acid to eicosapentaenoic acid. The inter-individual variability of SNPs in only 4 specific genes that had direct involvement in lutein response were ABCG2 (ATP binding cassette subfamily G member 2), ISX (intestine-specific homeobox), MTTP (microsomal triglyceride transfer protein) and ELOVL2 (Borel et al., 2014). ABCA1, APOA1 (apolipoprotein A1), APOB (apolipoprotein B), COBLL1 (cordon-bleu WH2 repeat protein-like 1), INSIG2 (insulin-induced gene 2), IRS1 (insulin receptor substrate 1), LIPC (lipase C, hepatic type), LPL (lipoprotein lipase), and MC4R (melanocortin 4 receptor) were the 9 genes out of the 29 genes that had a relation to postprandial chylomicron response (Desmarchelier et al., 2014). Reports suggested that the metabolism of lutein was also modulated by the APOE genotype (Huebbe, Lange, Lietz & Rimbach, 2016). While few genes encoding for subsequent proteins are involved in lutein absorption and transport, the mechanism is still not clear on the modulatory effect of variations in other genes on the bioavailability of lutein as per studies (Bhat and Mamatha, 2021). Irrespective of the current findings on the genetic factors that modulate the bioavailability of lutein, further research is needed to gain clarity and a better understanding of their mechanisms and outcomes in varying populations.

CONCLUSION

Lutein faces a combination of extrinsic challenges from source matrix and processing conditions prior to consumption, and intrinsic challenges from the host during gastrointestinal digestion. Both these categories of challenges are crucial in deciding the fate of lutein to be absorbed in the intestine. Although individual effects of the factors have been addressed in this chapter, the modulation of oral bioavailability of lutein is a resultant of combined phenomenon of these factors. In addition to the recent findings and advancements, further investigations on individual effects of processing conditions are required for explaining consequences of the factors that are currently unclear. There has been extensive research in the past decade in preparing delivery systems to accommodate lutein and improve its bioavailability. This chapter could also aid such studies which could utilize the processing conditions that favour oral bioavailability of lutein.

ACKNOWLEDGMENTS

The authors express their sincere gratitude to Prof. Dr. Praveen Kumar Shetty, Director (R&D), Nitte (DU) and Prof. Dr. Anirban Chakraborty, Director (NUCSER), Nitte (DU) for providing research facilities. The authors are grateful to Prof. Dr. Indrani Karunasagar, Director (DST-NUTEC), Nitte (DU) and Prof. Dr. Iddya Karunasagar, Advisor (Research and Patent), Nitte (DU) for their constant support and guidance.

REFERENCES

Abdel-Aal, El-Sayed M., J. Christopher Young, Humayoun Akhtar, and Iwona Rabalski. 2010. "Stability of lutein in wholegrain bakery products naturally high in lutein or fortified with free lutein." *Journal of agricultural and food chemistry* 58, no. 18: 10109-17. doi:10.1021/jf102400t.

Abdel-Aal, El-Sayed M., J. Christopher Young, Iwona Rabalski, Pierre Hucl, and Judith Fregeau-Reid. 2007. "Identification and quantification of seed carotenoids in selected wheat species." *Journal of agricultural and food chemistry* 55, no. 3: 787-94. doi:10.1021/jf062764p.

Amengual, Jaume, M. Airanthi K. Widjaja-Adhi, Susana Rodriguez-Santiago, Susanne Hessel, Marcin Golczak, Krzysztof Palczewski, and Johannes Von Lintig. 2013. "Two Carotenoid Oxygenases Contribute to Mammalian Provitamin A Metabolism." *Journal of Biological Chemistry* 288, no. 47: 34081-96. doi:10.1074/jbc.M113.501049.

Aruna, Goruspudi, Bangera Sheshappa Mamatha, and Vallikannan Baskaran. 2009. "Lutein content of selected Indian vegetables and vegetable oils determined by HPLC." *Journal of Food Composition and Analysis* 22, no. 7-8: 632-36. doi:10.1016/j.jfca.2009.03.006.

Baskaran, Villikannan, Tatsuya Sugawara, and Akihiko Nagao. 2003. "Phospholipids affect the intestinal absorption of carotenoids in mice." *Lipids* 38, no. 7: 705-11. doi:10.1007/s11745-003-1118-5.

Beta, Trust, and Taeyoung Hwang. 2018. "Influence of heat and moisture treatment on carotenoids, phenolic content, and antioxidant capacity of orange maize flour." *Food Chemistry* 246: 58-64. doi:10.1016/j.foodchem.2017.10.150.

Bhat, Ishani, and Bangera Sheshappa Mamatha. 2021. "Genetic factors involved in modulating lutein bioavailability." *Nutrition Research (New York, NY)* 91: 36-43. doi:10.1016/j.nutres.2021.04.007.

Bhat, Ishani, Gopinath Haripriya, Nishithkumar Jogi, and Bangera Sheshappa Mamatha. 2021. "Carotenoid composition of locally found seaweeds of Dakshina Kannada district in India." *Algal Research* 53: 102154. doi:10.1016/j.algal.2020.102154.

Bone, Richard. A., John. T. Landrum, Luis. Fernandez, and S. L. Tarsis. 1988. "Analysis of the macular pigment by HPLC: retinal distribution and age study." *Investigative Ophthalmology & Visual Science* 29, no. 6: 843-49.

Boon, Caitlin S., D. Julian McClements, Jochen Weiss, and Eric A. Decker. 2010. "Factors influencing the chemical stability of carotenoids in foods." *Critical Reviews in Food Science and Nutrition* 50, no. 6: 515-32. doi:10.1080/10408390802565889.

Borel, Patrick, Charles Desmarchelier, Marion Nowicki, Romain Bott, Sophie Morange, and Nathalie Lesavre. 2014. "Interindividual variability of lutein bioavailability in healthy men: characterization, genetic variants involved, and relation with fasting plasma lutein concentration." *The American Journal of Clinical Nutrition* 100, no. 1: 168-75. doi:10.3945/ajcn.114.085720.

Borel, Patrick, Fabien Szabo de Edelenyi, Stéphanie Vincent-Baudry, Christiane Malezet-Desmoulin, Alain Margotat, Bernard Lyan, Jean-Marie Gorrand, Nathalie Meunier, Sophie Drouault-Holowacz, and Severine Bieuvelet. 2011. "Genetic variants in BCMO1 and CD36 are associated with plasma lutein concentrations and macular pigment optical density in humans." *Annals of Medicine* 43, no. 1: 47-59. doi:10.3109/07853890.2010.531757.

Borel, Patrick, Georg Lietz, Aurélie Goncalves, Fabien Szabo de Edelenyi, Sophie Lecompte, Peter Curtis, Louisa Goumidi, Muriel J. Caslake, Elizabeth A. Miles, Christopher Packard, Phillip C. Calder, John C. Mathers, Anne M. Minihane, Franck Tourniaire, Emmanuelle Kesse-Guyot, Pilar Galan, Serge Hercberg, Christina Breidenassel, Marcela Gonzalez Gross, Myriam Moussa, Aline Meirhaeghe, and Emmanuelle Reboul. 2013. "CD36 and SR-BI are involved in cellular uptake of provitamin A carotenoids by Caco-2 and HEK cells, and some of their genetic variants are associated with plasma concentrations of these micronutrients in humans." *The Journal of Nutrition* 143, no. 4: 448-56. doi:10.3945/jn.112.172734.

Borel, Patrick, Pascal Grolier, M. Armand, Anne Partier, H. Lafont, D. Lairon, and Véronique Azaïs-Braesco. 1996. "Carotenoids in biological emulsions: solubility, surface-to-core distribution, and release from lipid droplets." *Journal of lipid research* 37, no. 2: 250-61. doi: 10.1016/S0022-2275(20)37613-6.

Burgos, Gabriela, Lupita Muñoa, Paola Sosa, Merideth Bonierbale, Thomas zum Felde, and Carlos Díaz. 2013. "In vitro bioaccessibility of lutein and zeaxanthin of yellow fleshed boiled potatoes." *Plant Foods for Human Nutrition* 68, no. 4: 385-90. doi: 10.1007/s11130-013-0381-x.

Carey, Martin C., and Donald M. Small. 1970. "The characteristics of mixed micellar solutions with particular reference to bile." *The American Journal of Medicine* 49, no. 5: 590-608. doi:10.1016/S0002-9343(70)80127-9.

Chen, Chun-Yen, Shih-Hsin Ho, Chen-Chun Liu, and Jo-Shu Chang. 2017. "Enhancing lutein production with Chlorella sorokiniana Mb-1 by optimizing acetate and nitrate concentrations under mixotrophic growth." *Journal of the Taiwan Institute of Chemical Engineers* 79: 88-96. doi:10.1016/j.jtice.2017.04.020.

Chung, Rosanna WS, Per Leanderson, Nelly Gustafsson, and Lena Jonasson. 2019. "Liberation of lutein from spinach: Effects of heating time, microwave-reheating and liquefaction." *Food Chemistry* 277: 573-78. doi:10.1016/j.foodchem.2018.11.023.

Corte-Real, Joana, Cédric Guignard, Manon Gantenbein, Bernard Weber, Kim Burgard, Lucien Hoffmann, Elke Richling, and Torsten Bohn. 2017. "No influence of supplemental dietary calcium intake on the bioavailability of spinach carotenoids in humans." *British Journal of Nutrition* 117, no. 11: 1560-69. doi:10.1017/S0007114517001532.

Corte-Real, Joana, Mohammed Iddir, Christos Soukoulis, Elke Richling, Lucien Hoffmann, and Torsten Bohn. 2016. "Effect of divalent minerals on the bioaccessibility of pure carotenoids and on physical properties of gastro-intestinal fluids." *Food Chemistry* 197: 546-53. doi:10.1016/j.foodchem.2015.10.075.

Dall'Osto, Luca, Stefano Cazzaniga, Mauro Bressan, David Paleček, Karel Židek, Krishna K. Niyogi, Graham R. Fleming, Donatas Zigmantas, and Roberto Bassi. 2017. "Two mechanisms for dissipation of excess light in monomeric and trimeric light-harvesting complexes." *Nature Plants* 3, no. 5: 1-9. doi:10.1038/nplants.2017.33.

Davidov-Pardo, Gabriel, Cansu Ekin Gumus, and David Julian McClements. 2016. "Lutein-enriched emulsion-based delivery systems: Influence of pH and temperature on physical and chemical stability." *Food Chemistry* 196: 821-27. doi:10.1016/j.foodchem. 2015.10.018.

de la Parra, Columba, Sergio O. Serna Saldivar, and Rui Hai Liu. 2007. "Effect of processing on the phytochemical profiles and antioxidant activity of corn for production of masa, tortillas, and tortilla chips." *Journal of Agricultural and Food Chemistry* 55, no. 10: 4177-83. doi:10.1021/jf063487p.

de Morais Cardoso, Leandro, Tatiana Aguiar Montini, Soraia Silva Pinheiro, Helena Maria Pinheiro-Sant'Ana, Hércia Stampini Duarte Martino, and Ana Vládia Bandeira Moreira. 2014. "Effects of processing with dry heat and wet heat on the antioxidant profile of sorghum." *Food Chemistry* 152: 210-17. doi:10.1016/j.foodchem. 2013.11.106.

dela Seña, Carlo, Sureshbabu Narayanasamy, Kenneth M. Riedl, Robert W. Curley, Steven J. Schwartz, and Earl H. Harrison. 2013. "Substrate specificity of purified recombinant human β-carotene 15, 15′-oxygenase (BCO1)." *Journal of Biological Chemistry* 288, no. 52: 37094-103. doi:10.1074/jbc.M113.507160.

Demmig-Adams, Barbara, and William W. Adams. 2002. "Antioxidants in photosynthesis and human nutrition." *Science* 298, no. 5601: 2149-53. doi:10.1126/science.1078002.

Desmarchelier, Charles, Jean-Charles Martin, Richard Planells, Marguerite Gastaldi, Marion Nowicki, Aurélie Goncalves, René Valéro, Denis Lairon, and Patrick Borel. 2014. "The postprandial chylomicron triacylglycerol response to dietary fat in healthy male adults is significantly explained by a combination of single nucleotide polymorphisms in genes involved in triacylglycerol metabolism." *The Journal of Clinical Endocrinology & Metabolism* 99, no. 3: E484-88. doi:10.1210/jc.2013-3686.

During, Alexandrine, Sundari Doraiswamy, and Earl H. Harrison. 2008. "Xanthophylls are preferentially taken up compared with β-carotene by retinal cells via a SRBI-dependent mechanism." *Journal of Lipid Research* 49, no. 8: 1715-24. doi:10.1194/jlr.M700580-JLR200.

Dwyer, James H., Mohamad Navab, Kathleen M. Dwyer, Kholood Hassan, Ping Sun, Anne Shircore, Susan Hama-Levy, Greg Hough, Xuping Wang, Thomas Drake, C. Noel Bairey Merz, and Alan M. Fogelman. 2001. "Oxygenated carotenoid lutein and progression of early atherosclerosis: the Los Angeles atherosclerosis study." *Circulation* 103, no. 24: 2922-27. doi:10.1161/01.CIR.103.24.2922.

Eisenhauer, Bronwyn, Sharon Natoli, Gerald Liew, and Victoria M. Flood. 2017. "Lutein and zeaxanthin—food sources, bioavailability and dietary variety in age-related macular degeneration protection." *Nutrients* 9, no. 2: 120: 1-14. doi: 10.3390/nu9020120.

Eriksen, Jane N., Amy Y. Luu, Lars O. Dragsted, and Eva Arrigoni. 2016. "In vitro liberation of carotenoids from spinach and Asia salads after different domestic kitchen procedures." *Food Chemistry* 203: 23-27. doi:10.1016/j.foodchem.2016.02.033.

Fatani, Amal J., Salim S. Al-Rejaie, Hatem M. Abuohashish, Abdullah Al-Assaf, Mihir Y. Parmar, and Mohammed M. Ahmed. 2015. "Lutein dietary supplementation attenuates streptozotocin-induced testicular damage and oxidative stress in diabetic rats." *BMC Complementary and Alternative Medicine* 15, no. 1: 1-10. doi:10.1186/s12906-015-0693-5.

Galasso, Christian, Cinzia Corinaldesi, and Clementina Sansone. 2017. "Carotenoids from marine organisms: Biological functions and industrial applications." *Antioxidants* 6, no. 4: 96. 1-33. doi:10.3390/antiox6040096.

Gleize, Béatrice, Franck Tourniaire, Laurence Depezay, Romain Bott, Marion Nowicki, Lionel Albino, Denis Lairon, Emmanuelle Kesse-Guyot, Galan Pilar, Hercberg Serge and Borel Patrick. 2013. "Effect of Type of TAG Fatty Acids on Lutein and Zeaxanthin Bioavailability." *British Journal of Nutrition* 110, no. 1: 1–10. doi:10.1017/S0007114512004813.

Goldberg, Jack, Gordon Flowerdew, Ellen Smith, Jacob A. Brody, and Mark OM Tso. 1988. "Factors associated with age-related macular degeneration: an analysis of data from the first National Health and Nutrition examination survey." *American Journal of Epidemiology* 128, no. 4: 700-10. doi:10.1093/oxfordjournals.aje.a115023.

Goñi, Isabel, José Serrano, and Fulgencio Saura-Calixto. 2006. "Bioaccessibility of β-carotene, lutein, and lycopene from fruits and vegetables." *Journal of Agricultural and Food Chemistry* 54, no. 15: 5382-87. doi:10.1021/jf0609835.

Gutiérrez-Uribe, Janet A., Carlos Rojas-García, Silverio García-Lara, and Sergio O. Serna-Saldivar. 2014. "Effects of lime-cooking on carotenoids present in masa and tortillas produced from different types of maize." *Cereal Chemistry* 91, no. 5: 508-12. doi:10.1094/CCHEM-07-13-0145-R.

Han, Hao, Wei Cui, Linzhi Wang, Yufang Xiong, Liegang Liu, Xiufa Sun, and Liping Hao. 2015. "Lutein prevents high fat diet-induced atherosclerosis in ApoE-deficient mice by inhibiting NADPH oxidase and increasing PPAR expression." *Lipids* 50, no. 3: 261-73. doi:10.1007/s11745-015-3992-1.

Handelman, Garry J., E. A. Dratz, C. C. Reay, and J. G. Van Kuijk. 1988. "Carotenoids in the human macula and whole retina." *Investigative Ophthalmology & Visual Science* 29, no. 6: 850-55.

Handelman, Garry J., Zachary D. Nightingale, Alice H. Lichtenstein, Ernst J. Schaefer, and Jeffrey B. Blumberg. 1999. "Lutein and zeaxanthin concentrations in plasma after dietary supplementation with egg yolk." *The American Journal of Clinical Nutrition* 70, no. 2: 247-51. doi:10.1093/ajcn.70.2.247.

Harrison, Earl H. 2019. "Mechanisms of transport and delivery of vitamin A and carotenoids to the retinal pigment epithelium." *Molecular Nutrition & Food Research* 63, no. 15: 1801046: 1-7. doi:10.1002/mnfr.201801046.

Heber, David, and Qing-Yi Lu. 2002. "Overview of mechanisms of action of lycopene." *Experimental Biology and Medicine* 227, no. 10: 920-23. doi:10.1177/153537020222701013.

Huang, Weiping, Yan Lin, Mingxia He, Yuhao Gong, and Junchao Huang. 2018. "Induced high-yield production of zeaxanthin, lutein, and β-carotene by a mutant of Chlorella zofingiensis." *Journal of Agricultural and Food Chemistry* 66, no. 4: 891-97. doi:10.1021/acs.jafc.7b05400.

Huang, Yang-Mu, Hong-Liang Dou, Fei-Fei Huang, Xian-Rong Xu, Zhi-Yong Zou, Xin-Rong Lu, and Xiao-Ming Lin. 2015. "Changes following supplementation with lutein and zeaxanthin in retinal function in eyes with early age-related macular degeneration: a randomised, double-blind, placebo-controlled trial." *British Journal of Ophthalmology* 99, no. 3 (2015a): 371-75. doi:10.1136/bjophthalmol-2014-305503.

Huang, Yang-Mu, Hong-Liang Dou, Fei-Fei Huang, Xian-Rong Xu, Zhi-Yong Zou, and Xiao-Ming Lin. 2015. "Effect of supplemental lutein and zeaxanthin on serum, macular pigmentation, and visual performance in patients with early age-related macular degeneration." *BioMed Research International* 2015 (2015b). 564738:1-8. doi:10.1155/2015/564738.

Huebbe, Patricia, Jennifer Lange, Georg Lietz, and Gerald Rimbach. 2016. "Dietary beta-carotene and lutein metabolism is modulated by the APOE genotype." *Biofactors* 42, no. 4: 388-96. doi:10.1002/biof.1284.

Humphries, Julia M., and Frederick Khachik. 2003. "Distribution of lutein, zeaxanthin, and related geometrical isomers in fruit, vegetables, wheat, and pasta products." *Journal of Agricultural and Food Chemistry* 51, no. 5: 1322-27. doi:10.1021/jf026073e.

Hwang, Eun-Sun, and Gun-Hee Kim. 2013. "Effects of various heating methods on glucosinolate, carotenoid and tocopherol concentrations in broccoli." *International Journal of Food Sciences and Nutrition* 64, no. 1: 103-11. doi:10.3109/09637486.2012.704904.

Iddir, Mohammed, Giulia Dingeo, Juan Felipe Porras Yaruro, Faiza Hammaz, Patrick Borel, Thomas Schleeh, Charles Desmarchelier, Yvan Larondelle, and Torsten Bohn. 2020. "Influence of soy and whey protein, gelatin and sodium caseinate on carotenoid bioaccessibility." *Food & Function* 11, no. 6: 5446-59. doi:10.1039/D0FO00888E.

Kamoshita, Mamoru, Eriko Toda, Hideto Osada, Toshio Narimatsu, Saori Kobayashi, Kazuo Tsubota, and Yoko Ozawa. 2016. "Lutein acts via multiple antioxidant pathways in the photo-stressed retina." *Scientific Reports* 6, no. 1: 1-10. doi:10.1038/srep30226.

Kean, Ellie G., Bruce R. Hamaker, and Mario G. Ferruzzi. 2008. "Carotenoid bioaccessibility from whole grain and degermed maize meal products." *Journal of Agricultural and Food Chemistry* 56, no. 21: 9918-26. doi:10.1021/jf8018613.

Khachik, Frederick, Christopher J. Spangler, J. Cecil Smith, Louise M. Canfield, Andrea Steck, and Hanspeter Pfander. 1997. "Identification, quantification, and relative concentrations of carotenoids and their metabolites in human milk and serum." *Analytical Chemistry* 69, no. 10: 1873-81. doi:10.1021/ac961085i.

Khachik, Frederick, Gary R. Beecher, Mudlagiri B. Goli, William R. Lusby, and James C. Smith Jr. 1992. "Separation and identification of carotenoids and their oxidation products in the extracts of human plasma." *Analytical Chemistry* 64, no. 18: 2111-22. doi:10.1021/ac00042a016.

Kijlstra, Aize, Sanne M. van der Made, Jogchum Plat, and Tos TJM Berendschot. 2017. "Lipoprotein changes following consumption of lutein-enriched eggs are associated with enhanced lutein bioavailability." *Journal of Food Nutrition & Research* 5: 362-69. doi:10.12691/jfnr-5-6-2.

Kishimoto, Yoshimi, Chie Taguchi, Emi Saita, Norie Suzuki-Sugihara, Hiroshi Nishiyama, Wei Wang, Yasunobu Masuda, and Kazuo Kondo. 2017. "Additional consumption of one egg per day increases serum lutein plus zeaxanthin concentration and lowers oxidized low-density lipoprotein in moderately hypercholesterolemic males." *Food Research International* 99: 944-49. doi:10.1016/j.foodres.2017.03.003.

Komuro, Marina, Naoki Shimizu, Ryo Onuma, Yurika Otoki, Junya Ito, Shunji Kato, Ohki Higuchi, Keiichi Sudo, Seiichiro Suzuki, Teruo Miyazawa, Takahiro Eitsuka, and Kiyotaka Nakagawa. 2017. "Analysis of Lutein in Mugwort (*Artemisia princeps* Pamp.) Paste and Evaluation of Manufacturing Processes." *Journal of Oleo Science* 66, no. 11: 1257-62. doi:10.5650/jos.ess17117.

Korobelnik, Jean-François, Marie-Bénédicte Rougier, Marie-Noëlle Delyfer, Alain Bron, Bénédicte MJ Merle, Hélène Savel, Geneviève Chêne, Cécile Delcourt, and Catherine Creuzot-Garcher. 2017. "Effect of dietary supplementation with lutein, zeaxanthin, and ω-3 on macular pigment: a randomized clinical trial." *JAMA Ophthalmology* 135, no. 11: 1259-66. doi:10.1001/jamaophthalmol.2017.3398.

Kostic, Dragana, Wendy S. White, and James A. Olson. 1995. "Intestinal absorption, serum clearance, and interactions between lutein and beta-carotene when administered to human adults in separate or combined oral doses." *The American Journal of Clinical Nutrition* 62, no. 3: 604-10. doi:10.1093/ajcn/62.3.604.

Krinsky, Norman I., John T. Landrum, and Richard A. Bone. 2003. "Biologic mechanisms of the protective role of lutein and zeaxanthin in the eye." *Annual Review of Nutrition* 23, no. 1: 171-201. doi:10.1146/annurev.nutr.23.011702.073307.

Landrum, John T., and Richard A. Bone. 2001 "Lutein, zeaxanthin, and the macular pigment." *Archives of Biochemistry and Biophysics* 385, no. 1: 28-40. doi:10.1006/abbi.2000.2171.

Lefsrud, Mark, Dean Kopsell, Adam Wenzel, and Joseph Sheehan. 2007. "Changes in kale (*Brassica oleracea* L. var. acephala) carotenoid and chlorophyll pigment concentrations during leaf ontogeny." *Scientia Horticulturae* 112, no. 2: 136-41. doi:10.1016/j.scienta.2006.12.026.

Ma, Tingting, Chengrui Tian, Jiyang Luo, Xiangyu Sun, Meiping Quan, Cuiping Zheng, and Jicheng Zhan. 2015. "Influence of technical processing units on the α-carotene, β-carotene and lutein contents of carrot (*Daucus carrot* L.) juice." *Journal of Functional Foods* 16: 104-13. doi:10.1016/j.jff.2015.04.020.

Maiani, Giuseppe, María Jesús Periago Castón, Giovina Catasta, Elisabetta Toti, Isabel Goñi Cambrodón, Anette Bysted, Fernando Granado-Lorencio, Begona Olmedilla-Alonso, Pia Knuthsen, Massimo Valoti, Volker Bohm, Esther Mayer-Miebach, Diana Behsnilian, and Ulrich Schlemmer. 2009. "Carotenoids: actual knowledge on food sources, intakes, stability and bioavailability and their protective role in humans." *Molecular Nutrition & Food Research* 53, no. S2: S194-S218. doi:10.1002/mnfr.200800053.

Mamatha, Bangera Sheshappa, and Vallikannan Baskaran. 2011. "Effect of micellar lipids, dietary fiber and β-carotene on lutein bioavailability in aged rats with lutein deficiency." *Nutrition* 27, no. 9: 960-66. doi:10.1016/j.nut.2010.10.011.

Mamatha, Bangera Sheshappa, Ranganathan Arunkumar, and Vallikannan Baskaran. 2012. "Effect of processing on major carotenoid levels in corn (Zea mays) and selected vegetables: Bioavailability of lutein and zeaxanthin from processed corn in mice." *Food and Bioprocess Technology* 5, no. 4: 1355-63. doi:10.1007/s11947-010-0403-8.

Mamatha, Bangera Sheshappa, Ravi Kumar Sangeetha, and Vallikannan Baskaran. 2011. "Provitamin-A and xanthophyll carotenoids in vegetables and food grains of nutritional and medicinal importance." *International Journal of Food Science & Technology* 46, no. 2: 315-23. doi:10.1111/j.1365-2621.2010.02481.x.

Mangels, Ann Reed, Joanne M. Holden, Gary R. Beecher, Michele R. Forman, and Elaine Lanza. 1993. "Carotenoid content of fruits and vegetables: an evaluation of analytic data." *Journal of the American Dietetic Association* 93, no. 3: 284-96. doi:10.1016/0002-8223(93)91553-3.

Margier, Marielle, Caroline Buffière, Pascale Goupy, Didier Remond, Charlotte Halimi, Catherine Caris-Veyrat, Patrick Borel, and Emmanuelle Reboul. 2018. "Opposite effects of the spinach food matrix on lutein bioaccessibility and intestinal uptake lead to unchanged bioavailability compared to pure lutein." *Molecular Nutrition & Food Research* 62, no. 11: 1800185. 1-22. doi:10.1002/mnfr.201800185.

Marriage, Barbara J., Jennifer A. Williams, Yong S. Choe, Kevin C. Maki, Mustafa Vurma, and Stephen J. DeMichele. 2017. "Mono-and diglycerides improve lutein absorption in healthy adults: a randomised, double-blind, cross-over, single-dose study." *British Journal of Nutrition* 118, no. 10: 813-21. doi:10.1017/S0007114517002963.

Mora-Gutierrez, A., R. Attaie, MT Núñez de González, Y. Jung, S. Woldesenbet, and S. A. Marquez. 2018. "Complexes of lutein with bovine and caprine caseins and their impact on lutein chemical stability in emulsion systems: Effect of arabinogalactan." *Journal of Dairy Science* 101, no. 1: 18-27. doi:10.3168/jds.2017-13105.

Moukarzel, Adib A., Riad A. Bejjani, and Florence N. Fares. 2009. "Xanthophylls and eye health of infants and adults." *Le Journal medical libanais. The Lebanese Medical Journal* 57, no. 4: 261-67.

Murillo, Ana Gabriela, Gregory H. Norris, Diana M. DiMarco, Siqi Hu, Yangchao Luo, Christofer N. Blesso, and Maria-Luz Fernandez. 2016. "A nano-emulsion of lutein is more effective than regular lutein in reducing cholesterol-induced liver injury in guinea pigs." *The FASEB Journal* 30: 913-2. doi:10.1096/fasebj.30.1_supplement.913.2.

Nidhi, Bhatiwada, Bangera Sheshappa Mamatha, and V. Baskaran. 2014. "Olive oil improves the intestinal absorption and bioavailability of lutein in lutein-deficient mice." *European Journal of Nutrition* 53, no. 1: 117-26. doi:10.1007/s00394-013-0507-9.

Niesor, Eric J., Evelyne Chaput, Jean-Luc Mary, Andreas Staempfli, Andreas Topp, Andrea Stauffer, Haiyan Wang, and Alexandre Durrwell. 2014. "Effect of compounds affecting ABCA1 expression and CETP activity on the HDL pathway involved in intestinal absorption of lutein and zeaxanthin." *Lipids* 49, no. 12: 1233-43. doi:10.1007/s11745-014-3958-8.

Nolan, John M., Ekaterina Loskutova, Alan N. Howard, Rachel Moran, Riona Mulcahy, Jim Stack, Maggie Bolger, J. Dennison, Akuffo, K. O., Owens, N. and Thurnam D. I. 2014. "Macular pigment, visual function, and macular disease among subjects with Alzheimer's disease: an exploratory study." *Journal of Alzheimer's Disease* 42, no. 4: 1191-202. doi: 10.3233/JAD-140507.

Nolasco, Emerson, Junsi Yang, Ozan Ciftci, Danh C. Vu, Sophie Alvarez, Sheila Purdum, and Kaustav Majumder. 2021. "Evaluating the effect of cooking and gastrointestinal digestion in modulating the bio-accessibility of different bioactive compounds of eggs." *Food Chemistry* 344: 128623. 1-9. doi:10.1016/j.foodchem.2020.128623.

O'Conell, Orla, Lisa Ryan, Laurie O'Sullivan, S. Aisling Aherne-Bruce, and Nora M. O'Brien. 2008. "Carotenoid micellarization varies greatly between individual and mixed vegetables with or without the addition of fat or fiber." *International Journal for Vitamin and Nutrition Research* 78, no. 45: 238-46. doi:10.1024/0300-9831.78.45.238.

Othman, Rashidi, Nur Hidayah Noh, Farah Ayuni Mohd Hatta, and Mohd Aizat Jamaludin. 2018. "Natural carotenoid pigments of 6 Chlorophyta freshwater green algae species." *Journal of Pharmacy and Nutrition Sciences* 8: 1-5.

Panova, Ina G., Marina A. Yakovleva, Alexander S. Tatikolov, A. S. Kononikhin, Tatiana B. Feldman, Rimma A. Poltavtseva, E. N. Nikolaev, Gennady T. Sukhikh, and Mikhail A. Ostrovsky. 2017. "Lutein and its oxidized forms in eye structures throughout prenatal human development." *Experimental Eye Research* 160: 31-7. doi:10.1016/j.exer.2017.04.008.

Parker, Robert S. 1996. "Absorption, metabolism, and transport of carotenoids." *The FASEB Journal* 10, no. 5: 542-51. doi:10.1096/fasebj.10.5.8621054.

Peng, Mei-Ling, Hui-Fang Chiu, Hsuan Chou, Hui-Ju Liao, Shyan-Tarng Chen, Yue-Ching Wong, You-Cheng Shen, Kamesh Venkatakrishnan, and Chin-Kun Wang. 2016. "Influence/impact of lutein complex (marigold flower and wolfberry) on visual function with early age-related macular degeneration subjects: a randomized clinical trial." *Journal of Functional Foods* 24: 122-30. doi:10.1016/j.jff.2016.04.006.

Perry, Alisa, Helen Rasmussen, and Elizabeth J. Johnson. 2009. "Xanthophyll (lutein, zeaxanthin) content in fruits, vegetables and corn and egg products." *Journal of Food Composition and Analysis* 22, no. 1: 9-15. doi:10.1016/j.jfca.2008.07.006.

Pumilia, Gloria, Morgan J. Cichon, Jessica L. Cooperstone, Daniele Giuffrida, Giacomo Dugo, and Steven J. Schwartz. 2014. "Changes in chlorophylls, chlorophyll degradation products and lutein in pistachio kernels (*Pistacia vera* L.) during roasting." *Food Research International* 65: 193-98. doi:10.1016/j.foodres.2014.05.047.

Qiu, Xiang, Dan-Hong Gao, Xiao Xiang, Yu-Fang Xiong, Teng-Shi Zhu, Lie-Gang Liu, Xiu-Fa Sun, and Li-Ping Hao. 2015. "Ameliorative effects of lutein on non-alcoholic fatty liver disease in rats." *World Journal of Gastroenterology: WJG* 21, no. 26: 8061-72. doi:10.3748/wjg.v21.i26.8061.

Ram, Shristi, Madhusree Mitra, Freny Shah, Sushma Rani Tirkey, and Sandhya Mishra. 2020. "Bacteria as an alternate biofactory for carotenoid production: A review of its applications, opportunities and challenges." *Journal of Functional Foods* 67: 103867. 1-13. doi:10.1016/j.jff.2020.103867.

Ranganathan, Arunkumar, Mamatha Bangera Sheshappa, and Vallikannan Baskaran. 2014. "Quality characteristics and lutein bioavailability from maize and vegetable-based health food." *Journal of dietary supplements* 11, no. 2: 131-44. doi:10.3109/19390211.2013.859208.

Read, Andrew, Amanda Wright, and El-Sayed M. Abdel-Aal. 2015. "In vitro bioaccessibility and monolayer uptake of lutein from wholegrain baked foods." *Food Chemistry* 174: 263-69. doi:10.1016/j.foodchem.2014.11.074.

Reboul, Emmanuelle, Lydia Abou, Céline Mikail, Odette Ghiringhelli, Marc André, Henri Portugal, Dominique Jourdheuil-Rahmani, Marie-Josèphe Amiot, Denis Lairon, and Patrick Borel. 2005. "Lutein transport by Caco-2 TC-7 cells occurs partly by a facilitated process involving the scavenger receptor class B type I (SR-BI)." *Biochemical Journal* 387, no. 2: 455-61. doi:10.1042/BJ20040554.

Reboul, Emmanuelle, Sinay Thap, Franck Tourniaire, Marc André, Christine Juhel, Sophie Morange, Marie-Josèphe Amiot, Denis Lairon, and Patrick Borel. 2007. "Differential effect of dietary antioxidant classes (carotenoids, polyphenols, vitamins C and E) on lutein absorption." *British Journal of Nutrition* 97, no. 3: 440-46. doi:10.1017/S0007114507352604.

Renzi-Hammond, Lisa M., Emily R. Bovier, Laura M. Fletcher, L. Stephen Miller, Catherine M. Mewborn, Cutter A. Lindbergh, Jeffrey H. Baxter, and Billy R. Hammond. 2017. "Effects of a lutein and zeaxanthin intervention on cognitive function: a randomized, double-masked, placebo-controlled trial of younger healthy adults." *Nutrients* 9, no. 11: 1246. 1-13. doi: 10.3390/nu9111246.

Riedl, Judith, Jakob Linseisen, Jürgen Hoffmann, and Günther Wolfram. 1999. "Some dietary fibers reduce the absorption of carotenoids in women." *The Journal of Nutrition* 129, no. 12: 2170-76. doi:10.1093/jn/129.12.2170.

Roodenburg, Annet JC, Rianne Leenen, Karin H. van het Hof, Jan A. Weststrate, and Lilian BM Tijburg. 2000. "Amount of fat in the diet affects bioavailability of lutein esters but not of α-carotene, β-carotene, and vitamin E in humans." *The American Journal of Clinical Nutrition* 71, no. 5: 1187-93. doi:10.1093/ajcn/71.5.1187.

Rosemary, Thomas, Abimannan Arulkumar, Sadayan Paramasivam, Alicia Mondragon-Portocarrero, and Jose Manuel Miranda. 2019. "Biochemical, micronutrient and physicochemical properties of the dried red seaweeds *Gracilaria edulis* and *Gracilaria corticata*." *Molecules* 24, no. 12: 2225. 1-14. doi:10.3390/molecules24122225.

Saini, Ramesh Kumar, and Young-Soo Keum. 2019. "Microbial platforms to produce commercially vital carotenoids at industrial scale: an updated review of critical issues." *Journal of Industrial Microbiology and Biotechnology* 46, no. 5: 657-74. doi:10.1007/s10295-018-2104-7.

Sánchez, Celia, Ana Beatriz Baranda, and Iñigo Martínez de Marañón. 2014. "The effect of high pressure and high temperature processing on carotenoids and chlorophylls content in some vegetables." *Food Chemistry* 163: 37-45. doi:10.1016/j.foodchem.2014.04.041.

Sato, Yuki, Risa Suzuki, Masaki Kobayashi, Shirou Itagaki, Takeshi Hirano, Toshihiro Noda, Satoshi Mizuno, Mitsuru Sugawara, and Ken Iseki. 2012. "Involvement of cholesterol membrane transporter Niemann-Pick C1-like 1 in the intestinal absorption of lutein." *Journal of Pharmacy & Pharmaceutical Sciences* 15, no. 2: 256-64. doi:10.18433/J38K56.

Schaeffer, Jonathan L., Juliusz K. Tyczkowski, Carmen R. Parkhurst, and Pat B. Hamilton. 1988. "Carotenoid composition of serum and egg yolks of hens fed diets varying in carotenoid composition." *Poultry Science* 67, no. 4: 608-14. doi:10.3382/ps.0670608.

Sharavana, Gurunathan, and Vallikannan Baskaran. 2017. "Lutein downregulates retinal vascular endothelial growth factor possibly via hypoxia inducible factor 1 alpha and X-box binding protein 1 expression in streptozotocin induced diabetic rats." *Journal of Functional Foods* 31: 97-103. doi:10.1016/j.jff.2017.01.023.

Shen, Ru, Senpei Yang, Guanghua Zhao, Qun Shen, and Xianmin Diao. 2015. "Identification of carotenoids in foxtail millet (*Setaria italica*) and the effects of cooking methods on carotenoid content." *Journal of Cereal Science* 61: 86-93. doi:10.1016/j.jcs.2014.10.009.

Shyam, Rajalekshmy, Aruna Gorusupudi, Kelly Nelson, Martin P. Horvath, and Paul S. Bernstein. 2017. "RPE65 has an additional function as the lutein to meso-zeaxanthin isomerase in the vertebrate eye." *Proceedings of the National Academy of Sciences* 114, no. 41: 10882-87. doi:10.1073/pnas.1706332114.

Stinco, Carla M., Justyna Szczepańska, Krystian Marszałek, Carlos A. Pinto, Rita S. Inácio, Paula Mapelli-Brahm, Francisco J. Barba, Jose M. Lorenzo, Jorge A. Saraiva, and Antonio J. Meléndez-Martínez. 2019. "Effect of high-pressure processing on carotenoids profile, colour, microbial and enzymatic stability of cloudy carrot juice." *Food Chemistry* 299: 125112. 1-7. doi:10.1016/j.foodchem.2019.125112.

Thomas, Sara E., and Earl H. Harrison. 2016. "Mechanisms of selective delivery of xanthophylls to retinal pigment epithelial cells by human lipoproteins." *Journal of Lipid Research* 57, no. 10: 1865-78. doi:10.1194/jlr.M070193.

van het Hof, Karin H., Clive E. West, Jan A. Weststrate, and Joseph GAJ Hautvast. 2000. "Dietary factors that affect the bioavailability of carotenoids." *The Journal of Nutrition* 130, no. 3: 503-6. doi:10.1093/jn/130.3.503.

Vishwanathan, Rohini, Matthew J. Kuchan, Sarbattama Sen, and Elizabeth J. Johnson. 2014. "Lutein and preterm infants with decreased concentrations of brain carotenoids." *Journal of Pediatric Gastroenterology and Nutrition* 59, no. 5: 659-65. doi: 10.1097/MPG. 0000000000000389.

West, C. E., and J. J. Castenmiller. 1998. "Quantification of the SLAMENGHI" factors for carotenoid bioavailability and bioconversion." *International Journal for Vitamin and Nutrition Research.* 68, no. 6: 371-77.

Wolf-Schnurrbusch, Ute EK, Martin S. Zinkernagel, Marion R. Munk, Andreas Ebneter, and Sebastian Wolf. 2015. "Oral lutein supplementation enhances macular pigment density and contrast sensitivity but not in combination with polyunsaturated fatty acids." *Investigative Ophthalmology & Visual Science* 56, no. 13: 8069-74. doi:10.1167/iovs.15-17586.

Xavier, Ana Augusta Odorissi, Adriana Zerlotti Mercadante, Juan Garrido-Fernandez, and Antonio Perez-Galvez. 2014. "Fat content affects bioaccessibility and efficiency of enzymatic hydrolysis of lutein esters added to milk and yogurt." *Food Research International* 65: 171-76. doi:10.1016/j.foodres.2014.06.016.

Xavier, Ana Augusta Odorissi, Irene Carvajal-Lérida, Juan Garrido-Fernández, and Antonio Pérez-Gálvez. 2018. "In vitro bioaccessibility of lutein from cupcakes fortified with a water-soluble lutein esters formulation." *Journal of Food Composition and Analysis* 68: 60-4. doi:10.1016/j.jfca.2017.01.015.

Xiao, Yibo, Xi He, Qi Ma, Yue Lu, Fan Bai, Junbiao Dai, and Qingyu Wu. 2018. "Photosynthetic accumulation of lutein in Auxenochlorella protothecoides after heterotrophic growth." *Marine Drugs* 16, no. 8: 283. 1-15. doi:10.3390/md16080283.

Yang, Cheng, Maike Fischer, Chris Kirby, Ronghua Liu, Honghui Zhu, Hua Zhang, Yuhuan Chen, Yong Sun, Lianfu Zhang, and Rong Tsao. 2018. "Bioaccessibility, cellular uptake and transport of luteins and assessment of their antioxidant activities." *Food Chemistry* 249: 66-76. doi:10.1016/j.foodchem.2017.12.055.

Yi, Jiang, Yuting Fan, Wallace Yokoyama, Yuzhu Zhang, and Liqing Zhao. 2016. "Characterization of milk proteins–lutein complexes and the impact on lutein chemical stability." *Food Chemistry* 200: 91-7. doi:10.1016/j.foodchem.2016.01.035.

Yonova-Doing, Ekaterina, Pirro G. Hysi, Cristina Venturini, Katie M. Williams, Abhishek Nag, Stephen Beatty, SH Melissa Liew, Clare E. Gilbert, and Christopher J. Hammond. 2013. "Candidate gene study of macular response to supplemental lutein and zeaxanthin." *Experimental Eye Research* 115: 172-77. doi:10.1016/j.exer.2013.07.020.

Zhang, Songhao, Jing Ji, Siqi Zhang, Chunfeng Guan, and Gang Wang. 2020. "Effects of three cooking methods on content changes and absorption efficiencies of carotenoids in maize." *Food & Function* 11, no. 1: 944-54. doi:10.1039/C9FO02622C.

In: What to Know about Carotenoids
Editor: Robert M. Albert

ISBN: 978-1-68507-105-9
© 2021 Nova Science Publishers, Inc.

Chapter 3

TECHNOLOGICAL APPROACHES TO THE PRODUCTION OF YEAST CAROTENOIDS

Lucielen Oliveira Santos[*], *Pedro Garcia Pereira Silva,
Daniel Prescendo Júnior, Tabita Veiga Dias Rodrigues,
Erika Carvalho Teixeira*
and Janaina Fernandes de Medeiros Burkert
Federal University of Rio Grande, School of Chemistry and Food,
Rio Grande, RS, Brazil

ABSTRACT

Carotenoid pigments, which represent the most widespread group of natural pigments, are yellow, orange and red lipophilic substances whose structures exhibit much diversity. They are synthesized by plants, algae and several microorganisms, such as bacteria, microalgae, filamentous

---

[*] Corresponding Author's E-mail: santoslucielen@gmail.com.

fungi and yeasts. Despite their wide distribution in nature, industrial commercialization of these pigments is mainly obtained by chemical synthesis. Biotechnological synthesis is an interesting alternative since it is natural and commercially competitive, based on consumers' concern over excessive use of chemical additives in food. Therefore, improving efficiency of carotenoid biosynthesis during fermentation/cultivation is crucial. New strategies to obtain these biomolecules through the optimization of biotechnological processes which use yeasts and the development of new approaches to downstream steps have been studied to make them become economically viable and competitive in the industrial market. Thus, this chapter aims at reviewing the main parameters of the production of yeast carotenoids and the downstream processes to obtain carotenogenic extracts. As a result, several aspects, such as the medium composition (carbon and nitrogen concentration, alternative raw materials), technological parameters (agitation/aeration, pH, temperature, fed-batch process, magnetic field application, light irradiation), genetic engineering, knowledge about carotenoid recovery and potential applications, were considered.

Keywords: biosynthesis, optimization, downstream steps, potential applications

1. Introduction

Carotenoids are abundant natural pigments, which have had about 1117 molecules isolated and characterized (Mussagy, Khan, and Kot 2021). Production of natural carotenoids has increased due to the growing market of healthier food that provide health benefits, based on consumers' concern for excessive use of chemical additives in food (Tang et al. 2019). Carotenoids are used in the production of cosmetics, food and textiles, which reach higher market value when they are prepared with natural pigments (Mishra, Varjani, and Varma 2019). Global demand increased 5.7% per year and may reach $2.0 billion by 2022 (Cipolatti et al. 2019; Silva et al. 2020b).

Biotechnological synthesis of carotenoids is an interesting alternative to obtain natural carotenoids. Several yeast species have been used for producing different types of carotenoids. However, the most commonly studied yeasts are *Phaffia rhodozyma* (Zhuang, Jiang, and Zhu 2020; Silva et al. 2020b), *Rhodotorula* sp. (Cipolatti et al. 2019; Tang et al. 2019), *Rhodosporidium* sp. (Dias et al. 2020; Pham et al. 2020), *Sporidiobolus* sp. (Cipolatti et al. 2019; Li et al. 2019) and *Pichia* sp. (Veiga-Crespo, Araya-Garay, and Villa 2018; Zhang et al. 2020). Advantages of the biotechnological process are the controlled production conditions and the use of low-cost substrates as feedstock, such as agro-industrial coproducts; it makes the process more industrially viable and adds value to the coproducts under use (Cipolatti et al. 2019; Dias et al. 2015; Mishra, Varjani, and Varma 2019).

Several studies have focused on parameters that may interfere with pigment bioproduction in relation to variety and quantity of carotenoids. They are composition of the culture medium (Rodrigues et al. 2019; Pereira et al. 2019; Cipolatti et al. 2019), temperature, agitation, pH (Dias et al. 2016, 2015; Kot et al. 2017; Liu et al. 2018; Silva et al. 2020a), magnetic field application (Silva et al. 2020b), genetic engineering of the microorganism (Ahuja et al. 2019; Jin et al. 2018; Li, Swofford, and Sinskey 2020; Usmani et al. 2020) and LED during cultivation (Manowattana et al. 2020).

Another very important issue in biotechnological carotenoid production is the extraction of compounds that are produced intracellularly, where rigidity of the cell wall limits their extraction. Thus, different techniques may be used for releasing intracellular compounds (Michelon et al. 2012; Lopes et al. 2017; Mussagy et al. 2019). This chapter aims at reviewing technological approaches to the production of yeast carotenoids, considering medium composition, technological parameters, genetic engineering, knowledge about carotenoid recovery and potential applications.

2. CAROTENOIDS

Carotenoids are pigments that are widely distributed in nature, responsible for orange, yellow and red colors found in photosynthetic tissues, fungi, bacteria, yeast, algae, animals and non-photosynthetic parts of plants, such as fruit, flowers, seeds, roots, vegetables, birds, fish and crustaceans (Cipolatti et al. 2019; Lopes et al. 2017; Massoud and Khosravi-Darani 2017). They are one of the most important groups of natural pigments, due to their wide distribution and structural and functional diversity. In addition, they have biological importance due to their photochemical, biochemical and antioxidant properties (Eman 2019; Mussagy et al. 2020).

Well-known for their use as natural dyes, carotenoids also have important biological functions as precursors of vitamin A and antioxidant activity. More than 1117 carotenoids have so far been identified in nature. However, only about 40 are found in a typical human diet, and 90% of carotenoids in the diet and in the human body are represented by β-carotene, α-carotene, lycopene, lutein and cryptoxanthin. Different carotenoids derive essentially from modifications in the base structure by cyclization of end groups and by introduction of oxygen functions, which give them their characteristic colors and antioxidant properties (Mussagy, Khan, and Kot 2021; Mussagy et al. 2019).

Most carotenoids are tetraterpenoids, the unique structure that determines their potential biological functions and activities. Due to unsaturation, carotenoids are sensitive to light, temperature, acidity and oxidation; they are hydrophobic compounds, lipophilic, insoluble in water and soluble in solvents, such as acetone, alcohol and chloroform (Mussagy et al. 2020; Massoud and Khosravi-Darani 2017; Kirti et al. 2014; Elsanhoty, Al-Turki, and El-Razik 2017). To obtain yellow, at least seven conjugated bonds are required. The more conjugated bonds, the larger the absorption bands and the longer the wavelengths; in this case, they become redder (Sen, Barrow, and Deshmukh 2019; Saini and Keum 2019).

It is known that carotenoids may be precursors of vitamin A, whose conversion is directly associated with specific carotenoids (α-carotene, β-cryptoxanthin and β-carotene) and occurs naturally by enzymatic reactions in the liver. High symmetry in their structure suggests that cleavage occurs at the central position of the polyene and produces two molecules of vitamin A (Greaves et al. 2012; Rodriguez-Amaya 2019). Basically, the structure of vitamin A (retinol) is half the one of β-carotene molecules, with a water molecule added at the end of the polyene chain. Consequently, β-carotene is the most potent vitamin A carotenoid (Rodriguez-Amaya 2019). Antioxidants are considered beneficial to human health because of their ability to act against reactive oxygen species. Free radicals are traditionally the major cause of age-related and degenerative diseases. This fact leads to the hypothesis that increased intake of antioxidants may reduce the risk of aging-related diseases (Frengova and Beshkova 2009; Kirti et al. 2014; Massoud and Khosravi-Darani 2017).

Compounds with conjugated double bonds have an antioxidant effect on scavenging free radicals. An example is carotenoids, which may have a beneficial effect on the body by scavenging and extinguishing radicals. The number of conjugated double bonds affects not only the color, but also the antioxidant capacity of carotenoids. In addition, antioxidant protection is provided by acyclic carotenoids, which have nine or more conjugated double bonds (Mussagy et al. 2019; Saini and Keum 2019).

Thus, carotenoids are of great importance, not only because they may be applied to food as natural colorants or vitamin A precursors, but also because, based on epidemiological studies, a positive link is suggested between high dietary intake and tissue concentrations of carotenoids and low risk of chronic and cardiovascular diseases and certain types of cancers (Rodriguez-Amaya 2019).

Biotechnological carotenoid production by microorganisms for industrial application has been a prominent subject in recent years, since most industrially used carotenoids are obtained chemically or by extraction from plants and algae. However, as the result of concern for the use of chemical additives in food, and the impact that these synthetic molecules have on human health, there has been growing interest in carotenoids obtained naturally by biotechnological processes (Valduga et al. 2009; Mata-Gómez et al. 2014; Mussagy, Khan, and Kot 2021).

Microbial carotenoids have become an area of intense study due to the ease of increasing production by cultivation conditions and genetic manipulation (Eman 2019; Jin et al. 2018). Production and profile of carotenoids depend on the microorganism, culture medium and operating conditions, such as temperature, pH, agitation, aeration rate and light. Moreover, carotenoid-producing microorganisms show rapid growth rate and may produce large quantities with no influence of environmental conditions. Growing conditions may also be controlled to ensure production of commercially important carotenoids and the use of low-cost substrates (Valduga et al. 2009; Mussagy et al. 2019; Mata-Gómez et al. 2014; Massoud and Khosravi-Darani 2017).

3. Carotenoid Biosynthesis

Carotenoids are generally tetraterpenoids (C40) formed by the union of eight isoprenoid units (C5), except in the central position, where junction occurs in the tail-to-tail direction; they are linked in such a way that the result in a symmetric and linear molecule. The basic acyclic C40 structure, which can be modified by hydrogenation, dehydrogenation, cyclization and oxidation, represents the most widespread group of pigments in nature (Mussagy, Khan, and Kot 2021; Tang et al. 2019).

Due to differences in carotenoid structures, food exhibits characteristic colors, which can vary from pale yellow to intense red; it is related to the series of conjugated double bonds, which allow absorption of light in the visible region of the pigments that act as a chromophore (Rapoport et al. 2021; Li, Swofford, and Sinskey 2020).

Mevalonic acid is the first specific precursor of terpenoid biosynthesis; after undergoing several reactions, it forms geranyl diphosphate (C10), farnesyl diphosphate (C15) and geranylgeranyl diphosphate (C20). Dimerization of two molecules of geranylgeranyl diphosphate forms phytoene, the first compound of forty carbons not colored yet. However, through a series of dehydrogenations, other carotenes are formed (Saini and Keum 2019). The biosynthesis of conversion into phytoene can be divided into three steps: 1) formation of isopentenyl pyrophosphate (IPP); 2) conversion of IPP into geranyl pyrophosphate (GGPP); and 3) condensation of two GGPP molecules to form phytoene, which is the precursor of most carotenoids (Figure 1) (Eman 2019; Mata-Gómez et al. 2014; Usmani et al. 2020).

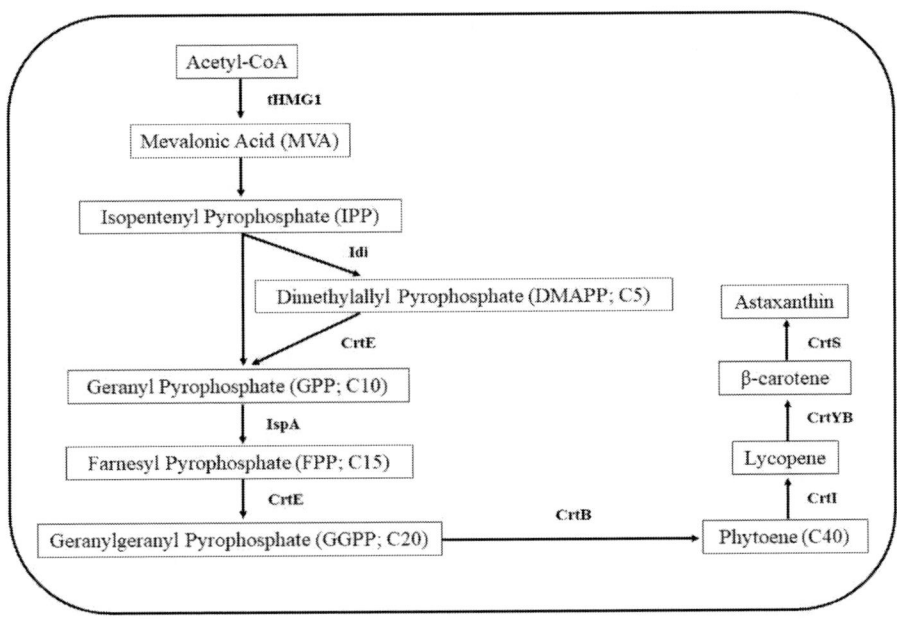

Figure 1. Pathway of yeast carotenoid, modified from Hara et al. (2014) and Saini and Keum (2019). The diagram consists of mevalonic acid and different carotenoid biosynthetic pathways. tHMG1, HMG-CoA reductase; Idi, Isopentenyl pyrophosphate isomerase; CrtE, Geranylgeranyl diphosphate synthase; IspA, Farnesyl diphosphate synthase; CrtB, Phytoene synthase; CrtI, Phytoene desaturase; CrtYB, Lycopene cyclase; CrtS, Astaxanthin synthase.

Substituents may be introduced with the formation of new conformation. Based on their structure, carotenoids can be divided into two major groups: (1) carotenes, which have only carbon and hydrogen in their structure and can be cyclized at one end of the molecule (phytoene, lycopene, α-carotene, and β-carotene); and (2) xanthophylls, which are carotenoid derivatives that have oxygen in their structure (lutein, canthaxanthin and astaxanthin) (Saini and Keum 2019; Rapoport et al. 2021).

Some carotenoids have biological functions as precursors of vitamin A. To carry out their function, they must have at least one β-ionone ring attached to an eleven-carbon polyene chain; β-carotene is the carotenoid that has the highest pro-vitamin A potential because it can be converted into two vitamin A molecules (Mussagy et al. 2019; Saini and Keum 2019). In yeasts, carotenoids are derived by sequence reactions through mevalonate biosynthesis. The main product 3-hydroxy-3-methylglutaryl-CoA (HMG-CoA) is reduced to mevalonic acid. This two-step reduction of HMG-CoA to mevalonate is highly controlled and a major controlling factor of their biosynthesis (Miao et al. 2010; Mussagy, Khan, and Kot 2021).

4. YEAST CAROTENOIDS

The ability of yeasts to grow in medium with high sugar content makes them industrially interesting. They are predominantly unicellular microorganisms with vegetative reproduction by budding or fission. Therefore, they have been studied in order to maximize and optimize carotenoid bioproduction in industries (Mata-Gómez et al. 2014; Rapoport et al. 2021).

Many microorganisms produce carotenoids, but not all have drawn the attention of industries. Yeasts have stood out because of their ability to grow on low-cost substrates and their high sugar content. Types of carotenoids and their relative amounts depend on the conditions of the culture medium, temperature, pH, aeration rate and light (Mata-Gómez et al. 2014; Eman 2019).

Carotenoid production by biotechnology can be performed with several yeast genera, such as *Rhodotorula, Rhodosporidium, Sporidiobolus, Sporobolomyces, Phaffia* and *Pichia*.

4.1. *Rhodotorula*

The genus *Rhodotorula* is a very heterogeneous group which comprises yeasts that have low fermentative capacity and do not form spores. It was discovered in 1927 by Harrison and is easily identified by the production of pigments in colonies due to the formation of natural carotenoid biosynthesis (Hernandez-Almanza et al. 2014; Tang et al. 2019).

This pigmentation is characteristic of the presence of carotenoids produced by the yeast to block certain electromagnetic wavelengths that would be harmful to the cell. The main carotenoids produced by *Rhodotorula* are β-carotene, torulene and torularhodin (Ribeiro et al. 2019; Frengova and Beshkova 2009).

Rhodotorula strains are polyphyletic, i.e., they contain *Rhodotorula* species that grow as single-celled yeast (monomorphic) and reproduce asexually by budding/fission (anamorphic). They also contain *Rhodosporidium* species that reproduce sexually (telemorphic) and alternate between a yeast phase and a dikaryotic filamentous one (dimorphic) (Morrow and Fraser 2009; Lyman et al. 2019; Tang et al. 2019).

Due to the biotechnological characteristics of yeasts that belong to the genus *Rhodotorula,* studies have been carried out to elucidate biotechnological processes that lead to maximum carotenoid production. In addition, these yeasts also produce several secondary metabolite products (saponifiable lipids for biodiesel production, antibacterial, organic acids, vitamins B and enzymes) (Hernandez-Almanza et al. 2014; Tang et al. 2019; Frengova and Beshkova 2009).

Production of carotenoid pigments takes place in isolates of the genus *Rhodotorula*, such as *Rhodotorula glutinis*, *Rhodotorula minuta*, *Rhodotorula mucilaginosa* and *Rhodosporidium toruloides.*

4.1.1. Rhodotorula glutinis

Rhodotorula glutinis is one of the species of the genus *Rhodotorula* which has wide distribution; colonies range from pink to red and asexual reproduction only occurs by budding (Buzzini and Martini 2000; Bhosale and Gadre 2002; Lyman et al. 2019). Most strains are spherical, ellipsoidal or elongated, aerobic and mesophilic, although some can thrive at low temperatures. Their colonies may be morphologically described as mucoid, smooth and shiny, while their cells are round and small. A notable advantage is that they use different carbon sources, such as glucose, galactose, sucrose, maltose, trehalose, lactose, ethanol, glycerol and lactose, to grow (Lyman et al. 2019; Kot et al. 2017; Aksu and Eren 2007; Mussagy et al. 2020). *R. glutinis* has a carotenoid profile composed of torularhodin, β-carotene (the main one), torulene, astaxanthin and lycopene. Regarding its industrial use, it also synthesizes numerous valuable compounds, such as lipids (single-cell oils) and enzymes (Valduga et al. 2009; Zhang, Zhang, and Tan 2014; Taskin et al. 2011; Tang et al. 2019).

4.1.2. Rhodotorula minuta

This worldwide species, whose colonies range from yellow to reddish, was discovered by Harrison in 1828. Concerning its morphology, its colonies are mucoid, smooth, shiny and red. Its cells are oval and small, with asexual reproduction by budding (Buzzini and Martini 2000; Bhosale 2004). *Rhodotorula minuta* has one of the most different patterns, i.e., the main carotenoid is β-zeacarotene, followed by phytofluene, in addition to β-carotene. Phytofluene is the colorless carotenoid that can only be found in R. *minuta*. Even though this yeast has low carotenoid contents, it may be used in food, pharmaceutical and cosmetic industries, since it consumes low-cost substrate (Maldonade, Rodriguez-Amaya, and Scamparini 2008; Silva et al. 2020c).

4.1.3. Rhodotorula mucilaginosa

In 1928, Harrison described *Rhodotorula mucilaginosa* (initially called *Rhodotorula rubra*) as a species that has several synonyms due to the variability of its strains regarding carbohydrate assimilation. Its colonies can be morphologically described as mucoid, smooth and opaque. Its color ranges from beige to orange and its cells are oval and large, with asexual reproduction by budding (Bhosale 2004; Frengova and Beshkova 2009; Valduga et al. 2009).

Almost all yeasts of the genus *Rhodotorula* produce characteristic carotenoids, such as β-carotene, torulene and torularhodin. Thus, *R.*

mucilaginosa is also a carotenoid-producing yeast that synthesizes β-carotene, lutein and torulene (Aksu and Eren 2005; Maldonade, Rodriguez-Amaya, and Scamparini 2008). Besides carotenoid, the red yeast has great potential to produce single cell oil (SCO). *R. mucilaginosa* has great potential due to its production of lipids and, as a red yeast, it can also synthesize carotenoids (Pereira et al. 2019).

4.1.4. Rhodosporidium toruloides

The yeast *Rhodosporidium toruloides*, which was first described by Banno in 1967, is the sexed state of *Rhodotorula glutinis*. It is known for its ability to produce neutral lipids and for being a red yeast (Zhu et al. 2012; Kot et al. 2016). Its colonies are red-orange due to the presence of carotenoids, mainly β-carotene, torulene and torularhodin (Bhosale 2004; Chandi and Gill 2011; Lyman et al. 2019). This yeast may grow on several carbon sources and its biomass is used in the production of antibiotics, synthesis of biofuels and the food industry (Lyman et al. 2019) *Rhodosporidium toruloides* is a nonpathogenic, aerobic and oleaginous red yeast that has been isolated from a variety of sources. It is able to accumulate lipids to more than 70% of its dry cell weight (Han et al. 2016; Zhu et al. 2012; Lyman et al. 2019).

4.2. Sporidiobolus pararoseus

Colonies of *Sporidiobolus pararoseus* can be described as creamy, smooth, shiny and red. Their cells are oval and medium-sized. *S. pararoseus* is a good source of lipids due to its high composition and to the synthesis of carotenoids (Han et al. 2016). It is able to produce carotenoids, such as lutein, β-cryptoxanthin and β-carotene, with an emphasis on β-carotene, torulene and torularhodin (Li et al. 2017, 2016; Otero et al. 2019). This yeast produces primary carotenoids β-carotene and torulene, as well as secondary carotenoids γ-carotene and torularhodin (Mannazzu et al. 2015; Mata-Gómez et al. 2014; Li et al. 2016).

4.3. *Sporobolomyces roseus*

Colonies of *Sporobolomyces roseus* can be described as mucoid, smooth, shiny and red. Their cells are oval and medium-sized. This yeast can be characterized by its production of β-carotene, γ-carotene, torulene and torularhodin (Valduga et al. 2009).

4.4. *Phaffia rhodozyma*

The yeast *Phaffia rhodozyma*, also known as *Xanthophyllomyces dendrorhous,* belongs to the basideomycetes class and has isolated elliptical vegetative cells, forming short chains or pairs. This yeast develops pink to red colonies and has a set of unique characteristics among yeast species (Libkind et al. 2011). *P. rhodozyma* may ferment sugars, such as glucose and xylose, and other sugars (such as sucrose), which can be obtained from starch, lignocellulosic materials, sugar beets and sugar cane (Frengova and Beshkova 2009; Rios et al. 2015).

Carotenoid synthesis by *P. rhodozyma* has advantages of simple nutritional requirements and short culture period (Silva et al. 2020b), with ability to produce intracellular carotenoids, such as astaxanthin, lutein and β-carotene, from microbial cultures (Michelon et al. 2012; Miao et al. 2010; Urnau et al. 2019). Astaxanthin is its main carotenoid (Urnau et al. 2018; Valduga et al. 2009; Frengova and Beshkova 2009; Libkind et al. 2011).

4.5. *Pichia fermentans*

There are not many studies of the yeast *Pichia fermentans*, but Otero isolated it in 2011 and observed its pink color in the colonies and production of carotenoids lutein, β-cryptoxanthin and β-carotene. Lutein is the carotenoid found at the highest concentration (Otero et al. 2019; Cipolatti et al. 2019).

Table 1 shows different carotenoid-producing yeasts.

Table 1. Carotenoid-producing yeasts

Microorganism	Main carotenoids	References
Rhodotorula glutinis sp.	Torularhodin, β-carotene, torulene, astaxanthin	(Valduga et al. 2009; Taskin et al. 2011; Zhang, Zhang, and Tan 2014; Mussagy et al. 2020, 2019)
Rhodotorula graminis sp.	Carotene, torulene	(Buzzini et al. 2005; Valduga et al. 2009)
Rhodotorula minuta sp.	β-carotene, β-zeacarotene, phytofluene	(Silva et al. 2020c; Maldonade, Rodriguez-Amaya, and Scamparini 2008)
Rhodotorula mucilaginosa sp.	Torulene, β-carotene, lutein	(Rodrigues et al. 2019; Cipolatti et al. 2019; Otero et al. 2019; Pereira et al. 2019)
Rhodosporidium toruloides sp.	Torulene, β-carotene, torularhodin	(Lyman et al. 2019; Massoud and Khosravi-Darani 2017)
Sporidiobolus pararoseus sp.	β-carotene, torulene, torularhodin	(Li et al. 2017, 2016; Otero et al. 2019)
Sporidiobolus salminolocor sp.	β-carotene	(Valduga et al. 2009)
Sporobolomyces roseus sp.	β-carotene, carotene, torulene, torularhodin	(Valduga et al. 2009)
Sporobolomyces ruberrimus sp.	Torularhodin, β-carotene	(Valduga et al. 2009)
Phaffia rhodozyma sp.	Astaxanthin, β-carotene, zeaxanthin, lycopene	(Frengova and Beshkova 2009; Libkind et al. 2011; Michelon et al. 2012; Silva et al. 2020b; Urnau et al. 2019)
Pichia fermentans sp.	Lutein, β-cryptoxanthin, β-carotene	(Otero et al. 2019; Cipolatti et al. 2019)

5. FACTORS THAT AFFECT PRODUCTION OF YEAST CAROTENOID

5.1. Culture Medium

Sources of carbon and nitrogen have been highlighted by studies of carotenoid production by biotechnological process. It occurs not only because of its influence on the development of microorganisms and on the synthesis of carotenoids, but also because it directly interferes with production costs (Dias et al. 2020; Borba et al. 2018). Yeasts accumulate carotenoids in response to environmental conditions and can grow on media with a high sugar content, making them industrially interesting. Some of them are *Rhodotorula glutinis* (Ribeiro et al. 2019), *Rhodotorula mucilaginosa* (Rodrigues et al. 2019), *Sporidiobolus pararoseus* (Cipolatti et al. 2019) and *Phaffia rhodozyma* (Rios et al. 2015). Glucose and sucrose are the most used carbon sources in production of yeast carotenoid. However, they may assimilate several carbon and nitrogen sources, including sugar cane molasses, glycerol, corn steep liquor and rice parboiling water, as shown in Table 2. Therefore, several studies have focused on the use of alternative sources of carbon and nitrogen (Borba et al. 2018; Kot et al. 2020; Rodrigues et al. 2019).

Rios et al. (2015) evaluated the use of rice parboiling water in the culture medium for carotenoid production by *Phaffia rhodozyma*. The culture was performed in shaken flasks (150 rpm) at 25 °C for 168 h. An experimental design was used for evaluating the composition of the medium: malt extract (1 to 16.25 g L^{-1}), peptone (1 to 16.25 g L^{-1}), sucrose (5 to 35 g L^{-1}) and rice parboiling water (4 to 162.5 g L^{-1}). Maximization of volumetric carotenoid production was achieved in the production medium with 16.25 g L^{-1} malt extract, 8.75 g L^{-1} peptone, 15 g L^{-1} sucrose and 87.5 g L^{-1} rice parboiling water in 144 h of cultivation. The result was 5300 µg L^{-1} (628.8 mg g^{-1}) of total carotenoids.

Dias et al. (2020) studied secondary brewery wastewater supplemented with sugarcane molasses which was used for lipid and carotenoid production by the yeast *Rhodosporidium toruloides* NCYC 92. Maximum carotenoid content (0.23 mg g^{-1}) was reached at 120 h of cultivation. Aksu and Eren (2007) employed glucose, molasses and lactose from cheese whey as carbon sources to produce carotenoid by *Rhodotorula glutinis*. The highest concentration (125 mg L^{-1}) was obtained with 20 g L^{-1} molasses, while the highest yield based on maximum cell concentration (35.5 mg g^{-1}) was obtained when 13.2 g L^{-1} whey lactose was added to the medium.

Table 2. Technological approach to production of yeast carotenoids

Yeast strain	Technological approach	Strategy	Carotenoid Production (µg L^{-1})	References
Rhodotorula mucilaginosa CCT 7688	Low-cost substrate	Sugar cane molasses and corn steep liquor	3726	(Rodrigues et al. 2019)
Sporidiobolus pararoseus CCT 7689	Low-cost substrate	Raw glycerol and corn steep liquor	634.5	(Cipolatti et al. 2019)
Rhodosporidium toruloides NCYC 921	Low-cost substrate	Brewery wastewater with sugarcane molasses and urea	1140	(Dias et al. 2020)
Sporidiobolus pararoseus KM 281507	Irradiation	Red light irradiation	5918	(Manowattana et al. 2020)
Rhodotorula glutinis CGMCC 2258	Irradiation	White light irradiation	2300	(Zhang, Zhang, and Tan 2014)
Xanthophyllomyces dendrorhous DSM 5626s	Irradiation	White light irradiation	950	(Stachowiak 2013)
Rhodotorula mucilaginosa BCRC 23454	pH	pH 5.0	2436.2	(Cheng and Yang 2016)
Rhodotorula glutinis LOCK R13	pH	pH 7.0	3489.8	(Kot et al. 2017)
Rhodosporidium toruloides NCYC 921	pH	pH 5.0	2781	(Dias et al. 2016)
Rhodotorula mucilaginosa BCRC 23454	Temperature	22 °C	2073.8	(Cheng and Yang 2016)
Xanthophyllomyces dendrorhous CBS 6938	Temperature	20 °C	874.1	(Liu et al. 2018)
Rhodotorula mucilaginosa CCT 3892	Temperature	22 °C	1130	(Silva et al. 2020a)
Phaffia rhodozyma NRRL Y-17268	Magnetic field	Intensity of 30 mT	1146.4	(Silva et al. 2020b)
Phaffia rhodozyma NRRL Y-17268	Magnetic field	Intensity of 30 mT	1184.6	(Prescendo Júnior et al. 2021)

Otero et al. (2019) selected three yeasts with high potential to produce carotenoid. They were collected and isolated in the southeastern Brazilian coast and identified as *Sporidiobolus pararoseus, Pichia fermentans and Rhodotorula mucilaginosa*. The use of two agro-industrial media, medium 1 (parboiled rice water and crude glycerol) and medium 2 (parboiled rice water and sugar cane molasses) were evaluated. The highest volumetric carotenoid concentration was achieved by *Sporidiobolus pararoseus*, which reached 820 and 710 μg L^{-1} in medium 1 and medium 2, respectively.

Rodrigues et al. (2019) studied agro-industrial media formulated only with co-products sugar cane molasses and corn steep liquor to produce carotenoids by the yeast *Rhodotorula mucilaginosa*. They successfully optimized the agro-industrial medium for carotenoid production; the yield of the optimized medium was higher than that of the medium whose carbon source was glucose. It shows that agro-industrial residues can replace traditional nitrogen and carbon sources in culture media.

5.2. Aeration and Agitation

Carotenogenesis is an aerobic process, in which oxygen is essential for substrate assimilation, cell growth and carotenoid synthesis, which is enhanced by a certain level of oxidative stress. Agitation is an indispensable process to make oxygen be more easily diffused into cells. Agitation and aeration are responsible for improving homogeneity of the culture medium and, consequently, for increasing nutrient availability (Mussagy et al. 2019; Frengova and Beshkova 2009; Mata-Gómez et al. 2014).

Borba et al. (2018) studied *Sporidiobolus pararoseus* and concluded that the best conditions to produce carotenoid are found in shake flaks with agitation at 150 rpm. Regarding *Phaffia rhodozyma*, positive effects were observed with agitation and aeration in carotenoid production, with increase of 33% in specific carotenoids bioproduction when 250 rpm and 1.5 vvm were used (Urnau et al. 2019). Aksu and Eren (2007) studied the aeration rate for 0, 1.2 and 2.4 vvm on growth and carotenoid production by *Rhodotorula glutinis*. The highest biomass and carotenoid concentrations were obtained at the highest aeration rate 2.4 vvm) by comparison with cultures without aeration.

Sharma and Ghoshal (2020) studied cultivation conditions (pH, temperature, agitation) and optimized carotenoid production by *Rhodotorula mucilaginosa* with the use of response methodology. They reached optimum conditions of pH 6.1 at 25.8 °C and 119.6 rpm. Aeration resulted in the best mixing of the cultivation medium, which is a crucial factor for growth and increase of approximately 100 mg g^{-1} when air was introduced into carotenoid production.

5.3. Medium pH

The pH of the medium is one of the key factors in biotechnological carotenoid production and its control is fundamental, since it affects yeast growth, metabolism, physiological activity and accumulation of pigment inside cells. Each yeast has an optimal pH value for its culture. Any change in the value leads to intracellular stress and activates the intracellular buffering of the medium. A pH value which is not ideal for a specific culture can affect extracytoplasmic enzymes, such as invertase, in addition to promoting changes in cell structure (Kot et al. 2017; Dias et al. 2015; Mussagy et al. 2019).

In the absence of a buffered medium, the pH value drops rapidly during the first stages of cultivation, a phenomenon that can be seen as a reflection of yeast activity, which absorbs amino acids, accumulates ions, excretes CO_2 in the medium or even excretes H^+ ions during ATP generation by respiration. ATP generation is linked to the gradient of protons generated by the difference between their extra and intracellular concentrations (Dias et al. 2016). Kot et al. (2017) observed complete growth inhibition of *Rhodotorula glutinis* in medium at pH 2.0, while significant reduction in growth was determined at pH 3.0. In the strongly acidified medium (pH 3.0), increase in torulene biosynthesis (56.2%) was determined and the amount of torularodine decreased significantly (4.5%). β-carotene content was similar to the value determined in the remaining media. Decrease in torularhodin biosynthesis by *Rhodotorula mucilaginosa* R-1 was also observed by Cheng and Yang (2016). Torularhodin content at initial pH 4.0 was 20.1%, while it increased to 36% in the medium at pH 7.0.

Many authors have reported that pH influences carotenoid production by different carotenogenic yeasts. Although Silva et al. (2020a) did not find considerable influence of the initial pH of the culture medium of *Rhodotorula mucilaginosa* on resulting biomass, it was essential for the synthesis of carotenoid within yeast cells. Reduction in pH from 7.0 to 5.0 favored the synthesis of these pigments. According to Cheng and Yang (2016), at initial pH 4, there is considerable reduction in biomass. On the other hand, at this pH, concentration of carotenoids is high, which means that, at low pH, yeast cells are forced to synthesize carotenoids. The disadvantage is that the volume of carotenoid is lower due to lower biomass concentration at this pH.

Dias et al. (2015) showed that the strategy of fed-batch cultivation with pH control in two stages (pH 4.0 in the growth phase; pH 5.0 in the lipid accumulation phase) significantly improved productivity of *Rhodosporidium toruloides* biomass and total carotenoid, by comparison with cultures in fed-batch conducted at fixed pH values of 5.5 and 4.0.

Thus, pH is an important factor that can affect biomass and production of yeast carotenoid. Controlling this factor is as important as controlling temperature, light, substrate source and other chemical and environmental factors, to obtain maximum carotenoid production on a large scale and lower costs of carotenoids produced by biotechnological route, by comparison with those produced chemically.

5.4. Temperature

Temperature influences both biomass production and carotenoid intracellular production. Most yeasts are mesophilic microorganisms, i.e., they need ideal moderate temperatures ranging from 20 to 45 °C. Cheng and Yang (2016) observed that the ideal temperature range for *Rhodotorula mucilaginosa* growth was between 20 and 30 °C. Below 20 °C, yeast growth was inhibited, while synthesis of carotenoid decreased when temperatures were above 30 °C. Besides, increase in temperature from 22 to 31 °C made percentages of β-carotene and torulene decrease from 29 to 23.9% and 50.5 to 32.6%, respectively, while torularhodin increased from 20.5% at 22 °C to 43.5% at 31°C.

Since enzyme concentration is controlled by temperature and carotenogenic syntheses are made by a multienzyme complex, temperature affects production of yeast carotenoid (Valduga et al. 2009; Silva et al. 2020a). The explanation for reduction carotenoid production in the study carried out by Cheng and Yang (2016) is the fact that temperatures above 30 °C can cause denaturation of enzymes involved in the process of carotenogenesis (Chandi and Gill 2011; Silva et al. 2020a).

A way to understand the effect of temperature change on production of yeast carotenoids from yeasts is the analysis of gene expression. Liu et al. (2018) performed an RT-PCR to compare transcriptional levels of genes involved in astaxanthin synthesis at different times of culture. Transcriptional levels of CrtE are the lowest carotenogenic genes, suggesting that CrtE might be an important factor that limits astaxanthin production. CrtE catalyzes the formation of geranylgeranyl diphosphate, which is generally considered the first step of the carotenoid pathway, because CrtE is part of the carotenogenic gene cluster in all microorganisms under investigation. High temperature showed its influence on the expression of CrtS and CrtI, which could be the reason for low astaxanthin production in the two-step process, because the enzyme CrtS has become a bottleneck of the carotenoid pathway. Therefore, further studies of omics tools are needed to better understand the influence of temperature on yeast genome. It has been known so far that temperature affects enzymes involved in the carotenogenic pathway.

5.5. Magnetic Field (MF)

In the literature, there is only one study that describes the influence of magnetic field (MF) on production of yeast carotenoid. (Silva et al. 2020b) observed the influence of 30 mT on the carotenogenic yeast *Phaffia rhodozyma*. The highest concentration of carotenoids was obtained with MF application throughout the cultivation, i.e., 1146.39 µg L^{-1} in 96 h. Thus, carotenoid production increased 59.4% by comparison with the control assay (without MF application).

MF causes cellular stress to yeast. Therefore, electro-activation of some enzymatic systems or metabolic pathways occurs positively and favors production of some intracellular compounds (Fologea et al. 1998). Their biosynthesis by Phaffia rhodozyma involves the formation of geranylgeranyl pyrophosphate (GGPP) and its conversion into phytoene, which is the precursor of most carotenoids, by the enzyme phytoene synthase (encoded by the CrtYB gene). Thus, the effect of MF application may influence carotenoid biosynthesis by electro-activation of the enzymatic system which involves phytoene synthase, linking two GGPP molecules tail to tail and producing phytoene (Silva et al. 2020b; Barredo et al. 2017).

5.6. Light Emitting Diode (LED)

The influence of Light Emitting Diode (LED) application on microbial cultivation has been studied to increase cell growth and/or production of compounds of interest. In carotenoid production by non-photosynthetic microorganisms, such as yeast, photo induction can be described by two aspects. Firstly, light plays a fundamental role on microorganism growth, as a stimulant. The second aspect considers that carotenoid accumulation in cells is associated with increase in the activity of enzymes involved in carotenoid biosynthesis (Bhosale 2004; Massoud and Khosravi-Darani 2017).

In general, light-induced oxidative and radiation damage can limit growth of some species of microorganisms, particularly those microorganisms that lack proper intracellular photoprotective substances, such as carotenoids. Carotenoids empower protection against light-induced damage and protect cells against the effect of singlet oxygen and excessive visible light and UV radiation (Hernandez-Almanza et al. 2014; Pham et al. 2020).

Sakaki et al. (2001) studied the effects of white light on growth and carotenoid production by *Rhodotorula glutinis* and reported that white light improved carotenoid yield, i.e., simultaneous increase in torularhodin and β-carotene was 180% and 14%, respectively. Zhang, Zhang, and Tan (2014) also reported that irradiation caused significant biomass improvement and carotenoid production enhancement of *Rhodotorula glutinis*. Some reports showed that LED derived irradiations at different colors may also promote carotenoid production and reach different results.

Red LED irradiation induced the highest β-carotene production by *Rhodotorula glutinis*, followed by blue, green and white LED, in this order (Yen and Yang 2012). Light intensity may also affect yeast. *Rhodotorula minuta* was found to tolerate up to 5000 Lux (Tada and Shiroishi 1990) whereas a mutant of *Rhodotorula glutinis* displayed poor growth when exposed to 1000 Lux (Bhosale and Gadre 2002).

Light irradiation was also considered an important factor to growth and synthesis of carotenoids by *Sporidiobolus pararoseus*. Manowattana et al. (2018) reported that growth, lipogenesis and carotenogenesis of *Sporidiobolus pararoseus* were inhibited in the absence of light. The effect of light irradiation by dark, natural, white, blue, green, yellow and red lights was studied by Manowattana et al. (2020). White light showed the highest yields of biomass, β-carotene and total carotenoids. Effects of the process carried out with blue and green light were similar to the one conducted with white light whereas carotenoid accumulation was lower when yellow and red light was used.

Concerning *Rhodosporidium toruloides*, carotenoid production was also found to be photo-inducible. The phenotype and gene expression analysis showed that *Rhodosporidium toruloides* responded to light (conditions by using 98 µmol m^{-2} s^{-1} photon on medium surface) by producing dark pigmentation with an associated increase in carotenoid production (Pham et al. 2020). Stachowiak (2013) studied astaxanthin production by *Xanthophyllomyces dendrorhous* and its mutants at illumination of 0 to 5000 Lux, which was close to natural light. The highest yields of astaxanthin were found in cultures at 600 Lux. Maximum pigment production by parental strain was 0.95 mg L^{-1} and 0.19 g kg^{-1} biomass, while by mutants, it was 1.23-1.51 mg L^{-1} and 0.34-0.39 g kg^{-1} biomass. Application of adequate illumination values in the range from 670 to 718 Lux to *Xanthophyllomyces dendrorhous* and light intensity of 5000 Lux turned out to be lethal for 80% strains under study.

These studies suggest that light is not a limiting factor in growth and carotenoid biosynthesis, but its intensity regulates the synthesis system and causes quantitative improvement in carotenoid produced by microorganisms. Light acts as a stressor and causes the cell to increase carotenoid production to protect itself.

6. Metabolic Engineering of Production of Yeast Carotenoid

In the biotechnology field, one of the alternatives to improve production and reduce costs of compounds of interest developed by microorganisms is the application of metabolic engineering techniques to obtain mutant strains (Li, Swofford, and Sinskey 2020). Thus, specific biochemical reactions may improve cellular properties by using the recombinant DNA technique (Mata-Gómez et al. 2014).

Regarding the use of mutant strains in carotenoid production, new approaches to molecular biology and metabolic engineering have led to cloning and manipulation of genes that are responsible for the biosynthesis and overproduction of pigments, thus, altering their molecular structure and color as desired. Cloning of genes that are responsible for carotenoid biosynthesis has been applied to carotenogenic and non-carotenogenic yeast cells (microbial vectors), making the process more economical and more reliable due to the adaptability of the microorganisms to mutagenesis (Table 3) (Sen, Barrow, and Deshmukh 2019; Jin et al. 2018; Ahuja et al. 2019).

Table 3. Metabolic engineering to produce yeast carotenoids

Metabolic engineering approach	Host organism	Major carotenoid	References
Expressed GGPP synthase, phytoene synthase, and phytoene desaturase	*Pichia pastoris* X-33	Lycopene	(Bhataya, Schmidt-Dannert, and Lee 2009)
Overexpressing the 3-hydroxy-3-methylglutaryl coenzyme A	*Saccharomyces cerevisiae* sp.	β-carotene	(Yan, Wen, and Duan 2012)
Novel combination of CrtZ-CrtW by mutagenesis	*Saccharomyces cerevisiae* sp	Astaxanthin	(Jin et al. 2018)
Overexpressing of CrtS	*Phaffia rhodozyma* sp.	Astaxanthin	(Miao et al. 2010)
Overexpressing the 3-hydroxy-3-methylglutaryl coenzyme A	*Rhodotorula mucilaginosa* KC8	β-carotene	(Wang et al. 2017)
Overexpressing of acetoacetyl-CoA thiolase, HMG-CoA synthase and HMG-CoA reductase	*Phaffia rhodozyma* sp.	β-carotene	(Hara et al. 2014)
Overexpressing of CrtZ	*Yarrowia lipolytica* sp.	Astaxanthin	(Tramontin et al. 2019)
Atmospheric and room temperature plasma (ARTP) mutagenesis	*Phaffia rhodozyma* Y1	Astaxanthin	(Zhuang, Jiang, and Zhu 2020)

The most common technique is random and selective mutagenesis. To apply it, physical and chemical factors are used to obtain specific biological characteristics in yeast mutants to overproduce carotenoids (Mata-Gómez et al. 2014; Jin et al. 2018). Another widely used technique is the manipulation of genes that are responsible for the carotenogenic pathway, in which two precursor enzymes, CrtW and CrtZ, are responsible different carotenoid intermediates and configure a varied profile of these compounds, such as β-carotene, astaxanthin, canthaxanthin and zeaxanthin (Ahuja et al. 2019; Usmani et al. 2020).

The engineered process of genetic mutation in yeast to produce carotenoids must be appropriate, i.e., it requires a characterized genetic system, physiology and regulation of its biological system. Thus, carotenoid precursor genes (from bacteria, yeasts and microalgae) have usually been applied to non-carotenogenic yeasts, such as *Candida. utilis* and *Saccharomyces cerevisiae* (Mata-Gómez et al. 2014; Sen, Barrow, and Deshmukh 2019).

Miura et al. (1998) studied the influence of different carotenogenic genes (CrtE, CrtB, CrtI, and CrtY) from different microorganisms applied to *Candida utilis*. They are responsible for carotenoid biosynthesis. The mutant yeast was able to produce carotenoids, such as β-carotene, lycopene and astaxanthin, efficiently. Liu et al. (2018) reported the use of mutagenic strategies to improve the CrtZ gene copy number adjustment from *Haematococcus pluvialis* to increase enzyme efficiency to act during the mevalonic pathway; they were able to produce astaxanthin by *Kluyveromyces marxianus* as the final product.

Regarding metabolic engineering with the use of the industrial yeast *Saccharomyces cerevisiae*, several studies have reported the use of expression techniques of carotenogenic genes extracted from the yeast *Phaffia rhodozyma*. The genes encoded CrtYB and CrtI are responsible for producing β-carotene through the bifunctional relation of these specific genes in the enzymatic activity during carotenoid biosynthesis (Zhou et al. 2015; Verwaal et al. 2007; Mata-Gómez et al. 2014).

Li et al. (2019) investigated overproduction of β-carotene by expressing 348 different genes from the mutant bacteria *Escherichia coli* through ultraviolet mutagenesis in *Sporidiobolus pararoseus*. The authors successfully expressed the carotenogenic gene CrtYB, responsible for β-carotene biosynthesis. Manowattana et al. (2020) studied random mutagenesis by applying physical and chemical factors, such as ethyl methane sulfonate combined with light irradiation, to enhance production of β-carotene by *Sporidiobolus pararoseus*.

Wang et al. (2017) studied the enhancement of carotenoid production by *Rhodotorula mucilaginosa* by using the atmospheric and room temperature plasma (ARTP) technique. The HMG1 gene was expressed through the mutant yeast, *Saccharomyces cerevisiae*, responsible for encoding 3-hydroxy-3-methylglutaryl coenzyme A (HMG-CoA) reductase during carotenoid biosynthesis. Pi et al. (2018) reported the use of genetic engineering tools, such as UV irradiation as a promoter of carotenogenic genes (CrtI, CrtE, CrtYB and tHMG1) from different yeasts expressed in the *Rhodotorula glutinis* genome for the overproduction of β-carotene.

Another study reports the optimization and selection of gene promoters of *Corynebacterium glutamicum* ATCC 13032s by using heterologous genes responsible for lycopene biosynthesis in the yeast *Pichia pastoris* GS115. By expanding the supply of Geranylgeranyl diphosphate (GGPP), both genes 3-hydroxy-3-methylglutaryl-coenzyme A (HMG CoA) and 3-hydroxy-3-methylglutaryl-CoA synthase (HMGS) were expressed within the mevalonate pathway to produce lycopene as the final product (Zhang et al. 2020).

Genetic engineering strategies are commonly used to produce β-carotene, astaxanthin and lycopene, since their biosynthesis has been widely reported by the literature. Thus, metabolic engineering and mutagenesis techniques have stood out in carotenoid production by carotenogenic and non-carotenogenic yeasts, due to their high added value by pharmaceutical and food industries.

7. DOWNSTREAM STEPS TO THE RECOVERY OF YEAST CAROTENOIDS

There are several natural pigments produced by microorganisms, but their commercialization is challenging. Their cost is five-fold the one of synthetic pigments. Besides, their collection, application, quality and regulation have also posed problems. Since microbial carotenoids are produced intracellularly, several downstream steps are required for their recovery and potential application. Thus, many studies have aimed to decrease the cost of the process and improve recovery yield (Mussagy et al. 2019; Michelon et al. 2012; Saini and Keum 2019).

Therefore, the development of alternative techniques of recovery of microbial carotenoids is necessary, since obtaining these biomolecules on a large scale is costly. Concerning yeast carotenoids, pretreatment of the biomass and cell disruption techniques by physical, chemical or enzymatic methods are needed, in order to obtain high yields (Table 4) (Mussagy et al. 2019; Mussagy, Khan, and Kot 2021; Mata-Gómez et al. 2014).

Table 4. Different methods of recovery of yeast carotenoids

Microrganism	Cell disruption method	Solvent extraction	Carotenoid content ($\mu g\ g^{-1}$)	References
Rhodotorula mucilaginosa	Physical (Ultrasonic bath)	Petroleum ether	193.5	(Lopes et al. 2017)
Rhodotorula glutinis	Chemical (Dimethyl sulfoxide)	Petroleum ether	647.5	(Taskin et al. 2011)
Phaffia rhodozyma	Enzymatic (β-1,3-glucanase)	Petroleum ether	156.9	(Michelon et al. 2012)
Xanthophyllomyces dendrorhous	Physical (Ultrasonic bath)	Methanol	614	(Urnau et al. 2018)
Phaffia rhodozyma	Chemical (Dimethyl sulfoxide)	Hexane	172.4	(Silva et al. 2020b)
Rhodotorula glutinis	Chemical (Protic ionic liquids)	Hexane/ethyl ether/acetic acid	~3000	(Mussagy et al. 2019)
Rhodotorula glutinis	Physical (Pulse eletric field)	Ethanol	375	(Martínez et al. 2018)

Prescendo Júnior et al. (2021) studied different pre-treatments of *Phaffia rhodozyma* biomass and different solvents for carotenoid recovery. Distinct drying and freezing techniques were used in the downstream step. The most effective pretreatment was drying and freezing for 24 h, followed by recovery with acetone and petroleum ether. Resulting volumetric concentration of carotenoids was around 1803.50 $\mu g\ L^{-1}$, i.e., it was 23.3% more effective than the standard procedure (drying and freezing for 48 h). Michelon et al. (2012) compared mechanical, chemical and enzymatic methods of extraction and recovery of carotenoids produced by *Phaffia rhodozyma*. The best result was obtained by the enzymatic lysis method, at pH 4.5, 55 °C, using β-1,3 glucanase with initial activity of 0.6 U mL^{-1} during 30 min, reaching 156.96 µg g^{-1} and extractability of 101.81%.

Lopes et al. (2017) compared the efficiency of different chemical and physical methods to extract carotenoids from both yeasts *Sporidiobolus pararoseus* and *Rhodotorula mucilaginosa*. Among the methods under study, ultrasonic bath and glass bead abrasion reached the best results for *Sporidiobolus pararoseus* (84.8 and 76.9 µg g^{-1}, respectively). In the case of the yeast *Rhodotorula mucilaginosa*, the most efficient method was the ultrasonic bath, which led to specific carotenoid concentration of around 193.5 µg g^{-1}. Martínez et al. (2018) studied the autolysis of *Rhodotorula glutinis* biomass using pulsed electric field-assisted extraction of carotenoids. Different conditions of the process were evaluated and maximum specific carotenoid concentration (375 µg g^{-1}) was reached after incubation for 24 h, at pH 8.0 and 25 °C. Subsequent extraction was carried out with ethanol.

Mussagy et al. (2019) studied the use of a new technology for carotenoid extraction by cell lysis with protic ionic liquids in *Rhodotorula glutinis* biomass. Twelve highly concentrated ammonium-based aqueous solutions of protic ionic liquids were used for evaluating efficiency of the extraction process. The most efficient extraction was performed at 65 °C with protic ionic liquids (concentration of 90%). Recoveries were 206.65, 112.8 and 17.21 µg mL^{-1} for β-carotene, torularodin and torulene, respectively. The same research group conducted by Mussagy et al. (2020) found a sustainable alternative for the recovery of carotenoids with the use of biosolvents (methanol, ethanol, ethyl acetate, isopropanol and cyclohexane). They found that, by using mixtures of ethyl acetate, ethanol and water (15/27/58%, v/v/v) in 3 extraction cycles, high efficiency was obtained in the process, i.e., around 75%, by comparison with conventional extraction methods (acetone, petroleum ether or hexane).

Downstream steps are critical to recover intracellular compounds since they affect recovery yield and carotenoid properties. Thus, it is important to develop alternative techniques to obtain natural pigments produced by yeasts with potential application to pharmaceutical and food industries.

8. APPLICATION OF YEAST CAROTENOIDS

The worldwide market of carotenoids will reach USD 1.15 billion in 2024 (it was USD 1.11 billion in 2019), according to a new study (More 2021). The Compound Annual Growth Rate (CAGR) for carotenoids, according to McWilliams (2018), was estimated at 5.7%. However, due to the crisis caused by the COVID-19 pandemic, the CAGR dropped drastically to 0.7% (More 2021), but it keeps having a significant world market.

According to More (2021), the major carotenoid demand comes from food, pharmaceutical and cosmetic industries, which mainly look for β-carotene, astaxanthin, canthaxanthin, lycopene and lutein due to their dye potential and health benefits.

Most commercial pigments are chemically synthesized in laboratories. In the 1960s, processes of large-scale production of some microbial carotenoids were developed but were later abandoned because they could not compete with synthetic manufacturing (Nelis and Leenheer 1991). Since the 90s, the "green wave" and public awareness of synthetic food additives have contributed to the revival of 'natural' carotenoid. As a result, industries have decreased their interests in synthetic molecules due to their toxic, carcinogenic and teratogenic properties and have given more attention to microbial sources as safe alternatives (Kirti et al. 2014; Nelis and Leenheer 1991).

Studies of carotenoids as pigment in foods have been reported by literature. Torularhodin, torulene and β-carotene extracted from *Rhodotorula mucilaginosa* were used as coloring agents for the production of sweet candy jelly by Elsanhoty, Al-Turki, and El-Razik (2017). Some products, such as Lucantin® Pink produced with astaxanthin by BASF, have been sold for aquaculture because they are efficient pigments to color some crustaceous, trout and salmon and make them more attractive in the market.

Vitamin A was the most important liposoluble vitamin discovered in the 1900s. After a few decades, a link between Vitamin A and carotenoids was discovered, and later on, many carotenoids were found to be metabolized by the body to form Vitamin A (Vandamme 1989; Greaves et al. 2012). Vitamin A performs many vital functions in humans, such as sight improvement, aid in growth and tooth and collagen formation. In addition, it is necessary for cell renewal and can be produced by the body from more than 50 carotenoids, but the most efficient one is β-carotene (Stahmann 2019; Kirti et al. 2014). It is the major carotenoid produced by the yeast *Blakeslea trispora* and corresponds to 15% of industrial β-carotene produced by microorganisms. This pigment is the most common food colorant (Stahmann 2019).

Astaxanthin is mainly produced by *Xanthophyllomyces dendrorhous*. It is applicable to egg production and as feed supplement for salmons, crabs, shrimps and chicken (Bampidis et al. 2019; Stahmann 2019). Astaxanthin also has health benefits in cardiovascular disease prevention, immune system boosting and cataract prevention due to its high antioxidant activity (Kirti et al. 2014). Furthermore, it might potentially reduce oxidative, inflammatory and coagulative stress (Medhi and Kalita 2021). Fouad et al. (2021) have recently stated that astaxanthin enhances breast anticancer activity.

Several studies have reported the role of carotenoids in modulating immunological reactions. It is caused by their antioxidant, anti-inflammatory and immunosuppressant properties (Kirti et al. 2014). Therefore, carotenoid properties can increase human immunity and avoid susceptibility to the SARS-Cov-2 (COVID-19) virus (Iddir et al. 2020; Jampilek and Kralova 2020; Medhi and Kalita 2021).

Since carotenoid is an important molecule to food, pharmaceutical and cosmetic products, it should be highlighted that carotenogenic yeasts can produce carotenoids by using different sources of carbon and nitrogen. One of the most studied substrates is agro-industrial residues, such as wastewater, which represents excessive amount of waste and high cost to industries. Thus, wastewater may be used as a yeast growth medium to produce carotenoids. The study carried out by Dias et al. (2020) was successful when it produced an amount of 33.43 mg L^{-1} carotenoid with the use of brewery effluent enriched with sugarcane molasses and urea as the growth medium for *Rhodosporidium toruloides*. Kot et al. (2020) produced 6.24 mg L^{-1} carotenoid with *Rhodotorula gracilis* cultivated with potato wastewater and glycerol.

CONCLUSION

The global market of carotenoids is in increasing demand as the result of their importance to food pigmentation and, consequently, to human and animal health. Sources of biotechnological routes are required to produce carotenoids since most of them have still been produced by chemical routes. Yeasts are potential carotenoid producers shown by the literature, mainly because of their capacity to produce carotenoid and biomass in different growth media, even in agro-industrial residues, to survive in many environmental cultivation conditions, to enable their cells to be easily disrupted for carotenoid extraction and to be an eco-friendly and natural route.

REFERENCES

Aksu, Z. and Tuğba, A. E. 2005. "Carotenoids production by the yeast *Rhodotorula mucilaginosa*: use of agricultural wastes as a carbon source." *Process Biochemistry* 40 (9): 2985–91. https://doi.org/10.1016/j.procbio.2005.01.011.

Aksu, Z. and Tuğba, A. E. 2007. "Production of carotenoids by the isolated yeast of *Rhodotorula glutinis*." *Biochemical Engineering Journal* 35 (2): 107–13. https://doi.org/10.1016/j.bej.2007.01.004.

Bampidis, V., Azimonti, G., Bastos, M. L., Christensen, H., Dusemund, B., Kouba, M., Durjava, M. K. Lópes-Alonso, M., Puente, S. L., Marcon, F., Mayo, B., Pechová, A., Petkova, M., Ramos, F., Sanz, Y., Villa, R.W., Woutersen, R., Bories, G., Brantom, P., Renshaw, D., Schlatter, J. R., Ackerl, R., Holcznecht, O., Steinkellner, H., Vettori, M. V. and Gropp, J. 2019. "Safety and efficacy of astaxanthin-dimethyldisuccinate (Carophyll® Stay-Pink 10%-CWS) for salmonids, crustaceans and other fish." *EFSA Journal* 17 (12). https://doi.org/10.2903/j.efsa.2019.5920.

Barredo, J. L., García-Estrada, C., Kosalkova, K. and Barreiro, C. 2017. "Biosynthesis of astaxanthin as a main carotenoid in the heterobasidiomycetous yeast *Xanthophyllomyces dendrorhous*." *Journal of Fungi* 3 (3). https://doi.org/10.3390/jof3030044.

Bhataya, A., Schmidt-Dannert, C. and Lee, P. C. 2009. "Metabolic engineering of *Pichia pastoris* X-33 for lycopene production." *Process Biochemistry* 44 (10): 1095–1102. https://doi.org/10.1016/j.procbio.2009.05.012.

Bhosale, P. 2004. "Environmental and cultural stimulants in the production of carotenoids from microorganisms." *Applied Microbiology and Biotechnology* 63 (4): 351–61. https://doi.org/10.1007/s00253-003-1441-1.

Bhosale, P. and Gadre, R. V. 2002. "Manipulation of temperature and illumination conditions for enhanced β-carotene production by mutant 32 of *Rhodotorula glutinis*." *Letters in Applied Microbiology* 34 (5): 349–53. https://doi.org/10.1046/j.1472-765X.2002.01095.x.

Borba, C. M., Tavares, M. N., Moraes, C. C. and Burkert, J. F. M. 2018. "Carotenoid production by *Sporidiobolus pararoseus* in agroindustrial medium: optimization of culture conditions in shake flasks and scale-up in a stirred tank fermenter." *Brazilian Journal of Chemical Engineering* 35 (2): 509–19. https://doi.org/10.1590/0104-6632.20180352s20160545.

Buzzini, P. and Martini, A. 2000. "Production of carotenoids by strains of *Rhodotorula glutinis* cultured in raw materials of agro-industrial origin." *Bioresource Technology* 71 (1): 41–44. https://doi.org/10.1016/S0960-8524(99)00056-5.

Buzzini, P., Martini, A., Gaetani, M., Turchetti, B., Pagnoni, U. M. and Davoli, P. 2005. "Optimization of carotenoid production by *Rhodotorula graminis* DBVPG 7021 as a function of trace element concentration by means of response surface analysis." *Enzyme and Microbial Technology* 36 (5–6): 687–92. https://doi.org/10.1016/j.enzmictec.2004.12.028.

Chandi, G. K. and Gill, B. S. 2011. "Production and characterization of microbial carotenoids as an alternative to synthetic colors: A review." *International Journal of Food Properties* 14 (3): 503–13. https://doi.org/10.1080/10942910903256956.

Cheng, Y. T. and Yang, C. F. 2016. "Using Strain *Rhodotorula mucilaginosa* to produce carotenoids using food wastes." *Journal of the Taiwan Institute of Chemical Engineers* 61 (April): 270–75. https://doi.org/10.1016/j.jtice.2015.12.027.

Cipolatti, E. P., Remedi, R. D., Sá, C. A., Rodrigues, A. B., Ramos, J. M. G., Burkert, C. A. V., Furlong, E. B. and Burkert, J. F. M. 2019. "Use of agroindustrial byproducts as substrate for production of carotenoids with antioxidant potential by wild yeasts." *Biocatalysis and Agricultural Biotechnology* 20 (May): 101208. https://doi.org/10.1016/j.bcab.2019.101208.

Dias, C., Reis, A., Santos J. A. L. and Silva, T. L. 2020. "Concomitant wastewater treatment with lipid and carotenoid production by the oleaginous yeast *Rhodosporidium toruloides* grown on brewery effluent enriched with sugarcane molasses and urea." *Process Biochemistry* 94 (January): 1–14. https://doi.org/10.1016/j.procbio.2020.03.015.

Dias, C., Silva, C., Freitas, C., Reis, A. and Silva, T. L. 2016. "Effect of medium ph on *Rhodosporidium toruloides* NCYC 921 carotenoid and lipid production evaluated by flow cytometry." *Applied Biochemistry and Biotechnology* 179 (5): 776–87. https://doi.org/10.1007/s12010-016-2030-y.

Dias, C., Sousa, S., Caldeira, J., Reis, A. and Silva, T. L. d. 2015. "New dual-stage ph control fed-batch cultivation strategy for the improvement of lipids and carotenoids production by the red yeast *Rhodosporidium toruloides* NCYC 921." *Bioresource Technology* 189: 309–18. https://doi.org/10.1016/j.biortech.2015.04.009.

Elsanhoty, R. M., Al-Turki, A. I. and El-Razik, M. M. A. 2017. "Production of carotenoids from *Rhodotorula mucilaginosa* and their applications as colorant agent in sweet candy." *Journal of Food, Agriculture and Environment* 15 (2): 21–26.

Eman, M. M. 2019. "Fungal and yeast carotenoids." *Journal of Yeast and Fungal Research* 10 (2): 30–44. https://doi.org/10.5897/jyfr2019.0192.

Fologea, D., Vassu-Dimov, T., Stoica, I., Csutak, O. and Radu, M. 1998. "Increase of *Saccharomyces cerevisiae* plating efficiency after treatment with bipolar electric pulses." *Bioelectrochemistry and Bioenergetics* 46 (2): 285–87. https://doi.org/10.1016/S0302-4598(98)00139-1.

Frengova, G. I. and Beshkova, D. M. 2009. "Carotenoids from *Rhodotorula* and *Phaffia*: Yeasts of biotechnological importance." *Journal of Industrial Microbiology and Biotechnology* 36 (2): 163–80. https://doi.org/10.1007/s10295-008-0492-9.

Greaves, R. F. 2012. "Vitamin A – Serum vitamin A analysis." In *Vitamin A and Carotenoids*, 162-83. Royal Society of Chemistry. https://doi.org/10.1039/9781849735506.

Han, M., Xu, Z. y., Du, C., Qian, H. and Zhang, W. G. 2016. "Effects of nitrogen on the lipid and carotenoid accumulation of oleaginous yeast *Sporidiobolus pararoseus*." *Bioprocess and Biosystems Engineering* 39 (9): 1425–33. https://doi.org/10.1007/s00449-016-1620-y.

Hara, K. Y., Morita, T., Mochizuki, M., Yamamoto, K., Ogino, C., Araki, M. and Kondo, A. 2014. "Development of a multi-gene expression system in *Xanthophyllomyces dendrorhous*." *Microbial Cell Factories* 13 (1). https://doi.org/10.1186/s12934-014-0175-3.

Hernandez-Almanza, A., Montanez, J. C., Aguilar-Gonzalez, M. A., Martmez-Avila, C., Rodriguez-Herrera, R. and Aguilar, C. N. 2014. "*Rhodotorula glutinis* as source of pigments and metabolites for food industry." *Food Bioscience* 5: 64–72. https://doi.org/10.1016/j.fbio. 2013.11.007.

Iddir, M., Brito, A., Dingeo, G., Campo, S. S. F., Samouda, H., Frano, M. R. and Bohn, T. 2020. "Strengthening the immune system and reducing inflammation and oxidative stress through diet and nutrition: considerations during the Covid-19 crisis." *Nutrients* 12 (6): 1–39. https://doi.org/10.3390/nu12061562.

Jampilek, J. and Kralova, K. 2020. "Potential of nanonutraceuticals in increasing immunity." *Nanomaterials* 10 (11): 1–47. https://doi.org/10.3390/nano10112224.

Jin, J., Wang, Y., Yao, M., Gu, X., Li, B., Liu, H., Ding, M., Xiao, W. and Yuan, Y. 2018. "Astaxanthin overproduction in yeast by strain engineering and new gene target uncovering." *Biotechnology for Biofuels* 11 (1): 1–15. https://doi.org/10.1186/s13068-018-1227-4.

Kirti, K., Amita, S., Priti, S., Kumar, A. M. and Jyoti, S. 2014. "Colorful world of microbes: carotenoids and their applications." *Advances in Biology* 2014: 1–13. https://doi.org/10.1155/2014/837891.

Kot, A. M., Błażejak, S., Kurcz, A., Bryś, J., Gientka, I., Bzducha-Wróbel, A., Maliszewska, M. and Reczek, L. 2017. "Effect of initial ph of medium with potato wastewater and glycerol on protein, lipid and carotenoid biosynthesis by *Rhodotorula glutinis*." *Electronic Journal of Biotechnology* 27 (May): 25–31. https://doi.org/10.1016/j.ejbt.2017.01.007.

Kot, A. M., Błażejak, S., Kurcz, A., Gientka, I. and Kieliszek, M. 2016. "*Rhodotorula glutinis*—Potential source of lipids, carotenoids, and enzymes for use in industries." *Applied Microbiology and Biotechnology* 100 (14): 6103–17. https://doi.org/10.1007/s00253-016-7611-8.

Kot, A. M., Błażejak, S., Kieliszek, M., Gientka, I. and Piwowarek, K. 2020. "Production of lipids and carotenoids by *Rhodotorula gracilis* ATCC 10788 yeast in a bioreactor using low - cost wastes." *Biocatalysis and Agricultural Biotechnology* 26 (May): 101634. https://doi.org/10.1016/j.bcab.2020.101634.

Li, C., Swofford, C. A. and Sinskey, A. J. 2020. "Modular engineering for microbial production of carotenoids." *Metabolic Engineering Communications* 10 (October 2019): e00118. https://doi.org/10.1016/j.mec.2019.e00118.

Li, C., Li, B., Zhang, N., Wang, Q., Wang, W. and Zou, H. 2019. "Comparative transcriptome analysis revealed the improved β-carotene production in *Sporidiobolus pararoseus* yellow mutant MuY9." *Journal of General and Applied Microbiology* 65 (3): 121–28. https://doi.org/10.2323/jgam.2018.07.002.

Li, C., Zhang, N., Li, B., Xu, Q., Song, J., Wei, N., Wang, W. and Zou, H. 2017. "Increased torulene accumulation in red yeast *Sporidiobolus pararoseus* NGR as stress response to high salt conditions." *Food Chemistry* 237: 1041–47. https://doi.org/10.1016/j.foodchem.2017.06.033.

Li, C., Zhang, N., Song, J., Wei, N., Li, B., Zou, H. and Han, X. 2016. "A single desaturase gene from red yeast *Sporidiobolus pararoseus* is responsible for both four- and five-step dehydrogenation of phytoene." *Gene* 590 (1): 169–76. https://doi.org/10.1016/j.gene.2016.06.042.

Libkind, D., Tognetti, C., Ruffini, A., Sampaio, J. P. and Broock, M. V. 2011. "*Xanthophyllomyces dendrorhous* (*Phaffia rhodozyma*) on stromata of cyttaria hariotii in northwestern patagonian nothofagus forests." *Revista Argentina de Microbiologia* 43 (3): 226–32. https://doi.org/10.1590/S0325-75412011000300011. [*Argentine Journal of Microbiology*]

Liu, S., Liu, B., Wang, H., Xiao, S., Li, Y. and Wang, J. 2018. "Production of astaxanthin at moderate temperature in *Xanthophyllomyces dendrorhous* using a two-step process." *Engineering in Life Sciences* 18 (10): 706–10. https://doi.org/10.1002/elsc.201800065.

Lopes, N. A., Remedi, R. D., Sá, C. S., Burkert, C. A. V. and Burkert, J. F. M. 2017. "Different cell disruption methods for obtaining carotenoids by *Sporodiobolus pararoseus* and *Rhodothorula mucilaginosa*." *Food Science and Biotechnology* 26 (3): 759–66. https://doi.org/10.1007/s10068-017-0098-y.

Lyman, M., Urbin, S., Strout, C. and Rubinfeld, B. 2019. "The oleaginous red yeast *Rhodotorula/Rhodosporidium* : A factory for industrial bioproducts." *Yeasts in Biotechnology*, 0–19. https://doi.org/10.5772/intechopen.84129.

Maldonade, I. R., Rodriguez-Amaya, D. B. and Scamparini, A. R. P. 2008. "Carotenoids of yeasts isolated from the brazilian ecosystem." *Food Chemistry* 107 (1): 145–50. https://doi.org/10.1016/j.foodchem.2007.07.075.

Manmeet, A., Jayesh, V., Mansi, V., Piyush, S., Harikrishna, R. and Vidhya, R.. 2019. "Astaxanthin: Current advances in metabolic engineering of the carotenoid." *High Value Fermentation Products* 1: 381–99. https://doi.org/10.1002/9781119460053.ch17.

Mannazzu, I., Landolfo, S., Silva, T. L. and Buzzini, P. 2015. "Red yeasts and carotenoid production: outlining a future for non-conventional yeasts of biotechnological interest." *World Journal of Microbiology and Biotechnology* 31 (11): 1665–73. https://doi.org/10.1007/s11274-015-1927-x.

Manowattana, A., Techapun, C., Laokuldilok, T., Phimolsiripol, Y. and Chaiyaso, T. 2020. "Enhancement of β-carotene-rich carotenoid production by a mutant S*poridiobolus pararoseus* and stabilization of its antioxidant activity by microencapsulation." *Journal of Food Processing and Preservation* 44 (8): 1–12. https://doi.org/10.1111/jfpp.14596.

Manowattana, A., Techapun, C., Watanabe, M. and Chaiyaso, T. 2018. "Bioconversion of biodiesel-derived crude glycerol into lipids and carotenoids by an oleaginous red yeast *Sporidiobolus Pararoseus* KM281507 in an airlift bioreactor." *Journal of Bioscience and Bioengineering* 125 (1): 59–66. https://doi.org/10.1016/j.jbiosc.2017.07.014.

Martínez, J. M., Delso, C., Angulo, J., Álvarez, I. and Raso, J. 2018. "Pulsed electric field-assisted extraction of carotenoids from fresh biomass of *Rhodotorula glutinis*." *Innovative Food Science and Emerging Technologies* 47 (April): 421–27. https://doi.org/10.1016/j.ifset.2018.04.012.

Massoud, R. and Khosravi-Darani, K. 2017. "A review on the impacts of process variables on microbial production of carotenoid pigments." *Food Biosynthesis*. Elsevier Inc. https://doi.org/10.1016/b978-0-12-811372-1.00006-3.

Mata-Gómez, L., Montañez, J., Méndez-Zavala, A. and Aguilar, C. 2014. "Biotechnological production of carotenoids by yeasts: An overview." *Microbial Cell Factories* 13 (1): 12. https://doi.org/10.1186/1475-2859-13-12.

McWilliams, A. 2018. *The global market for carotenoids*. https://www.bccresearch.com/market-research/food-and-beverage/the-global-market-for-carotenoids.html.

Medhi, J. and Kalita, M. C. 2021. "Astaxanthin: An algae-based natural compound with a potential role in human health-promoting effect: An updated comprehensive review." *Journal of Applied Biology and Biotechnology* 9 (1): 114–23. https://doi.org/10.7324/JABB.2021.9115.

Miao, L., Wang, Y., Chi, S., Yan, J., Guan, G., Hui, B. and Li, Y. 2010. "Reduction of fatty acid flux results in enhancement of astaxanthin synthesis in a mutant strain of *Phaffia rhodozyma*." *Journal of Industrial Microbiology and Biotechnology* 37 (6): 595–602. https://doi.org/10.1007/s10295-010-0706-9.

Michelon, M., Borba, T. M., Rafael, R. S., Burkert, C. A. V. and Burkert, J. F. M. 2012. "Extraction of carotenoids from *Phaffia rhodozyma*: A comparison between different techniques of cell disruption." *Food Science and Biotechnology* 21 (1): 1–8. https://doi.org/10.1007/s10068-012-0001-9.

Mishra, B., Varjani, S. and Varma, G. K. S. 2019. "Agro-industrial by-products in the synthesis of food grade microbial pigments: an eco-friendly alternative." In *Green Bio-Processes*, 245–65. Springer Singapore. https://doi.org/10.1007/978-981-13-3263-0_13.

Miura, Y., Kondo, K., Saito, T., Shimada, H., Fraser, P. D. and Misawa, N. 1998. "Production of the carotenoids lycopene, β-carotene, and astaxanthin in the food yeast *Candida utilis*." *Applied and*

Environmental Microbiology 64 (4): 1226–29. https://doi.org/10.1128/aem.64.4.1226-1229.1998.

More, A. 2021. "Carotenoids market 2021: Analysis of key trends, industry dynamics and future growth 2024 with top countries data." 360 Research Reports.

Morrow, C. A. and Fraser, J. A. 2009. "Sexual reproduction and dimorphism in the pathogenic basidiomycetes." *FEMS Yeast Research* 9 (2): 161–77. https://doi.org/10.1111/j.1567-1364.2008.00475.x.

Mussagy, C. U., Santos-Ebinuma, V. C., Gonzalez-Miquel, M., Coutinho, J. A. P. and Pereira, J. F. B. 2019. "Protic ionic liquids as cell-disrupting agents for the recovery of intracellular carotenoids from yeast *Rhodotorula glutinis* CCT-2186." *ACS Sustainable Chemistry and Engineering* 7 (19): 16765–76. https://doi.org/10.1021/acssuschemeng.9b04247.

Mussagy, C. U., Santos-Ebinuma, V. C., Kurnia, K. A., Dias, A. C. R. V., Carvalho, P., Coutinho, J. A. P. and Pereira, J. F. B. 2020. "Integrative platform for the selective recovery of intracellular carotenoids and lipids from: *Rhodotorula glutinis* CCT-2186 yeast using mixtures of bio-based solvents." *Green Chemistry* 22 (23): 8478–94. https://doi.org/10.1039/d0gc02992k.

Mussagy, C. U., Khan, S. and Kot, A. M. 2021. "Current developments on the application of microbial carotenoids as an alternative to synthetic pigments." *Critical Reviews in Food Science and Nutrition* 0 (0): 1–15. https://doi.org/10.1080/10408398.2021.1908222.

Mussagy, C. U., Winterburn, J., Santos-Ebinuma, V. C. and Pereira, J. F. B. 2019. "Production and extraction of carotenoids produced by microorganisms." *Applied Microbiology and Biotechnology* 103 (3): 1095–1114. https://doi.org/10.1007/s00253-018-9557-5.

Nelis, H. J. and Leenheer, A. P.. 1991. "Microbial sources of carotenoid pigments used in foods and feeds." *Journal of Applied Bacteriology* 70 (3): 181–91. https://doi.org/10.1111/j.1365-2672.1991.tb02922.x.

Otero, D. M., Bulsing, B. A., Huerta, K. M., Rosa, C. A., Zambiazi, R. C., Burkert, C. A. V. and Burkert, J. F. M. 2019. "Carotenoid-producing yeasts in the brazilian biodiversity: isolation, identification and cultivation in agroindustrial waste." *Brazilian Journal of Chemical Engineering* 36 (1): 117–29. https://doi.org/10.1590/0104-6632. 20190361s20170433.

Pereira, R. N., Silveira, J. M., Burkert, J. F. M., Ores, J. C. and Burkert, C. A. V. 2019. "Simultaneous lipid and carotenoid production by stepwise fed-batch cultivation of *Rhodotorula mucilaginosa* with crude glycerol." *Brazilian Journal of Chemical Engineering* 36 (3): 1099–1108. https://doi.org/10.1590/0104-6632.20190363s20190199.

Pham, K. D., Shida, Y., Miyata, A., Takamizawa, T., Suzuki, Y., Ara, S., Yamazaki, H., Mesaki, K., Mori, K., Aburatani, S., Hirakawa, H., Tashiro, K., Kuhara, S., Takaku, H. and Ogasawara, W. 2020. "Effect of light on carotenoid and lipid production in the oleaginous yeast *Rhodosporidium toruloides*." *Bioscience, Biotechnology and Biochemistry* 84 (7): 1501–12. https://doi.org/10.1080/09168451. 2020.1740581.

Pi, H. W., Anandharaj, M., Kao, Y. Y., Lin, Y. J., Chang, J. J. and Li, W. H. 2018. "Engineering the oleaginous red yeast *Rhodotorula glutinis* for simultaneous β-carotene and cellulase production." *Scientific Reports* 8 (1): 2–11. https://doi.org/10.1038/s41598-018-29194-z.

Prescendo Júnior, D., Silva, P. G. P., Burkert, J. F. M. and Santos, L. O. 2021. "Improvement of downstream step for the extraction of carotenoids produced by the yeast *Phaffia rhodozyma*." *Brazilian Journal of Agrotechnology* 11 (2): 71–76. https://doi.org/10.18378/ rebagro.v12i2.8765.

Rapoport, A., Guzhova, I., Bernetti, L., Buzzini, P., Kieliszek, M. and Kot, A. M. 2021. "Carotenoids and some other pigments from fungi and yeasts." *Metabolites* 11 (2): 1–17. https://doi.org/10.3390/metabo 11020092.

Ribeiro, J. E. S., Sant'Ana, A. M. S., Martini, M., Sorce, C., Andreucci, A., Melo, D. J. N. and Silva, F. L. H. 2019. "*Rhodotorula glutinis* cultivation on cassava wastewater for carotenoids and fatty acids generation." *Biocatalysis and Agricultural Biotechnology* 22 (October). https://doi.org/10.1016/j.bcab.2019.101419.

Rios, D. A. S., Borba, T. M., Kalil, S. J. and Burkert, J. F. M. 2015. "Rice parboiling wastewater in the maximization of carotenoids bioproduction by *Phaffia rhodozyma*." *Science and Agrotechnology* 39 (4): 401–10. https://doi.org/10.1590/s1413-70542015000400011.

Rodrigues, T. V. D., Amore, T. D., Teixeira, E. C. and Burkert, J. F. M. 2019. "Carotenoid production by *Rhodotorula mucilaginosa* in batch and fed-batch fermentation using agroindustrial byproducts." *Food Technology and Biotechnology* 57 (3): 388–98. https://doi.org/10.17113/ftb.57.03.19.6068.

Rodriguez-Amaya, D. B. 2019. "Update on natural food pigments - A mini-review on carotenoids, anthocyanins, and betalains." *Food Research International* 124 (2017): 200–205. https://doi.org/10.1016/j.foodres.2018.05.028.

Saini, R. K. and Keum, Y. S. 2019. "Microbial platforms to produce commercially vital carotenoids at industrial scale: an updated review of critical issues." *Journal of Industrial Microbiology and Biotechnology* 46 (5): 657–74. https://doi.org/10.1007/s10295-018-2104-7.

Sakaki, H., Nakanishi, T., Tada, A., Miki, W. and Komemushi, S. 2001. "Activation of torularhodin production by *Rhodotorula glutinis* using weak white light irradiation." *Journal of Bioscience and Bioengineering* 92 (3): 294–97. https://doi.org/10.1016/S1389-1723(01)80265-6.

Sen, T., Barrow, C. J. and Deshmukh, S. K. 2019. "Microbial pigments in the food industry—Challenges and the way forward." *Frontiers in Nutrition* 6 (March): 1–14. https://doi.org/10.3389/fnut.2019.00007.

Sharma, R. and Ghoshal, G. 2020. "Optimization of carotenoids production by *Rhodotorula mucilaginosa* (MTCC-1403) using agro-industrial waste in bioreactor: A statistical approach." *Biotechnology Reports* 25: e00407. https://doi.org/10.1016/j.btre.2019.e00407.

Silva, J., Silva, F. L. H., Ribeiro, J. S. E., Melo, D. J. N., Santos, F.A. and Medeiros, L. L. 2020. "Effect of supplementation, temperature and ph on carotenoids and lipids production by *Rhodotorula mucilaginosa* on sisal bagasse hydrolyzate." *Biocatalysis and Agricultural*

Biotechnology 30 (September): 101847. https://doi.org/10.1016/j.bcab. 2020.101847.

Silva, P. G. P., Prescendo Júnior, D., Sala, L., Burkert, J. F. M. and Santos, L. O. 2020. "Magnetic field as a trigger of carotenoid production by *Phaffia rhodozyma*." *Process Biochemistry* 98: 131–38. https://doi.org/10.1016/j.procbio.2020.08.001.

Silva, S. R. S., Stamford, T. C. M., Albuquerque, W. W. C., Vidal, E. E. and Stamford, T. L. M. 2020. "Reutilization of residual glycerin for the produce β-carotene by *Rhodotorula minuta*." *Biotechnology Letters* 42 (3): 437–43. https://doi.org/10.1007/s10529-020-02790-8.

Stachowiak, B. 2013. "Effect of illumination intensities on astaxanthin synthesis by *Xanthophyllomyces dendrorhous* and its mutants." *Food Science and Biotechnology* 22 (4): 1033–38. https://doi.org/10.1007/s10068-013-0180-z.

Stahmann, K. P. 2019. "Vitamins and vitamin-like compounds: microbial production." *Encyclopedia of Microbiology*, no. October 2016: 569–80. https://doi.org/10.1016/B978-0-12-809633-8.13017-1.

Tada, M. and Shiroishi, M. 1990. "Mechanism of photoregulated carotenogenesis in *Rhodotorula minuta* VIII. Effect of mevinolin on photoinduced carotenogenesis." *Plant and Cell Physiology* 31 (3): 319–23. https://doi.org/10.1093/oxfordjournals.pcp.a077910.

Tang, W., Wang, Y., Zhang, J., Cai, Y. and He, Z. 2019. "Biosynthetic pathway of carotenoids in *Rhodotorula* and strategies for enhanced their production." *Journal of Microbiology and Biotechnology* 29 (4): 507–17. https://doi.org/10.4014/jmb.1901.01022.

Taskin, M., Sisman, T., Erdal, S and Kurbanoglu, E. B. 2011. "Use of waste chicken feathers as peptone for production of carotenoids in submerged culture of *Rhodotorula glutinis* MT-5." *European Food Research and Technology* 233 (4): 657–65. https://doi.org/10.1007/s00217-011-1561-2.

Tramontin, L. R. R., Kildegaard, K. R., Sudarsan, S. and Borodina, I. 2019. "Enhancement of astaxanthin biosynthesis in oleaginous yeast *Yarrowia lipolytica* via microalgal pathway." *Microorganisms* 7 (10). https://doi.org/10.3390/microorganisms7100472.

Urnau, L., Colet, R., Reato, P. T., Burkert, J. F. M., Rodrigues, E., Gomes, R., Jacques, R. A., Valduga, E. and Steffens, C. 2019. "Use of low-cost agro-industrial substrate to obtain carotenoids from *Phaffia rhodozyma* in a bioreactor." *Industrial Biotechnology* 15 (1): 25–34. https://doi.org/10.1089/ind.2018.0027.

Urnau, L., Colet, R., Soares, V. F., Franceschi, E., Valduga, E. and Steffens, C. 2018. "Extraction of carotenoids from *Xanthophyllomyces dendrorhous* using ultrasound-assisted and chemical cell disruption methods." *Canadian Journal of Chemical Engineering* 96 (6): 1377–81. https://doi.org/10.1002/cjce.23046.

Usmani, Z., Sharma, M., Sudheer, S., Gupta, V. K. and Bhat, R. 2020. "Engineered microbes for pigment production using waste biomass." *Current Genomics* 21 (2): 80–95. https://doi.org/10.2174/1389202921999200330152007.

Valduga, E., Tatsch, P. O., Tiggemann, L., Treichel, H., Toniazzo, G., Zeni, J., Luccio, M. D. and Fúrigo Júnior, A. 2009. "Carotenoid production: Microorganisms as a source of natural pigments." *New Chemistry* 32 (9): 2429–36. https://doi.org/10.1590/S0100-40422009000900036.

Vandamme, E. J. 1989. *Biotechnology of Vitamins, Pigments and Growth Factors*. Edited by Erick J. Vandamme. Dordrecht: Springer Netherlands. https://doi.org/10.1007/978-94-009-1111-6.

Veiga-Crespo, P., Araya-Garay, J. M. and Villa, T.G. 2018. "Engineering *Pichia pastoris* for the production of carotenoids." *Methods in Molecular Biology* 1852: 311–26. https://doi.org/10.1007/978-1-4939-8742-9_19.

Verwaal, R., Wang, J., Meijnen, J. P., Visser, H., Sandmann, G., Berg, J. A. V. D. and Ooyen, A. J. J. V. 2007. "High-level production of beta-carotene in *Saccharomyces cerevisiae* by successive transformation with carotenogenic genes from *Xanthophyllomyces dendrorhous*." *Applied and Environmental Microbiology* 73 (13): 4342–50. https://doi.org/10.1128/AEM.02759-06.

Wang, Q., Liu, D., Yang, Q. and Wang, P. 2017. "Enhancing carotenoid production in *Rhodotorula mucilaginosa* KC8 by combining mutation and metabolic engineering." *Annals of Microbiology* 67 (6): 425–31. https://doi.org/10.1007/s13213-017-1274-2.

Yan, G. L., Wen, K. R. and Duan, C. Q. 2012. "Enhancement of β-carotene Production by over-expression of hmg-coa reductase coupled with addition of ergosterol biosynthesis inhibitors in recombinant *Saccharomyces cerevisiae*." *Current Microbiology* 64 (2): 159–63. https://doi.org/10.1007/s00284-011-0044-9.

Yen, H. W. and Yang, Y. C. 2012. "The effects of irradiation and microfiltration on the cells growing and total lipids production in the cultivation of *Rhodotorula glutinis*." *Bioresource Technology* 107: 539–41. https://doi.org/10.1016/j.biortech.2011.12.134.

Zhang, X., Wang, D., Duan, Y., Zheng, X., Lin, Y. and Liang, S. 2020. "Production of lycopene by metabolically engineered *Pichia pastoris*." *Bioscience, Biotechnology and Biochemistry* 84 (3): 463–70. https://doi.org/10.1080/09168451.2019.1693250.

Zhang, Z., Zhang, X. and Tan. T. 2014. "Lipid and carotenoid production by *Rhodotorula glutinis* under irradiation/high-temperature and dark/low-temperature cultivation." *Bioresource Technology* 157: 149–53. https://doi.org/10.1016/j.biortech.2014.01.039.

Zhou, P., Ye, L., Xie, W., Lv, X. and Yu, H. 2015. "Highly efficient biosynthesis of astaxanthin in *Saccharomyces cerevisiae* by integration and tuning of algal CrtZ and Bkt." *Applied Microbiology and Biotechnology* 99 (20): 8419–28. https://doi.org/10.1007/s00253-015-6791-y.

Zhu, Z., Zhang, S., Liu, H., Shen, H., Lin, X., Yang, F., Zhou, Y. J. Jin, G., Ye, M., Zou, H. and Zhao, Z. K. 2012. "A multi-omic map of the lipid-producing yeast *Rhodosporidium toruloides*." *Nature Communications* 3. https://doi.org/10.1038/ncomms2112.

Zhuang, Y., Jiang, G. L. and Zhu, M. J. 2020. "Atmospheric and room temperature plasma mutagenesis and astaxanthin production from sugarcane bagasse hydrolysate by *Phaffia rhodozyma* mutant Y1." *Process Biochemistry* 91 (December 2019): 330–38. https://doi.org/10.1016/j.procbio.2020.01.003.

In: What to Know about Carotenoids
Editor: Robert M. Albert

ISBN: 978-1-68507-105-9
© 2021 Nova Science Publishers, Inc.

Chapter 4

ROSACEAE FRUITS AS A VALUABLE SOURCE OF CAROTENOIDS. A REVIEW

T. Negreanu-Pirjol[1], D. R. Popoviciu[2] and B. S. Negreanu-Pirjol[1]*

[1]Faculty of Pharmacy, "Ovidius" University of Constanta, Romania
[2]Faculty of Natural Sciences and Agricultural Sciences, "Ovidius" University of Constanta, Romania

ABSTRACT

Rosaceae is one of the largest families of flowering plants, comprising numerous species of culinary, medicinal and ornamental interest, featuring a wide variety of fruit types: pomes, drupes, polydrupes or polyachenes. This paper summarizes known data on total carotenoid contents and chemical diversity of Rosaceae fruits. Overall carotenoid content within this family is extremely variable, with some commonly grown fruits (apples, pears, cherries, strawberries) ranking among the lowest, while tropical *Rubus* species, followed by species in the *Sorbus*,

---

* Corresponding Author's E-mail: ticuta_np@yahoo.com.

Cotoneaster, *Pyracantha* or *Rosa* genera having remarkably high carotenoid contents. Among the known carotenoid compounds found in Rosaceae fruits are α-, β- and γ-carotene, β-cryptoxanthin, zeaxanthin, violaxanthin, rubixanthin, flavoxanthin, neoxanthin, lycoxanthin, capsanthin, xantophyll esters, lutein, lutein epoxide, lycopene, prolycopene, β-citraurin, capsorubin etc., with variations among different species. Also, carotenoid content shows significant variations between fruit skins and flesh, with skins having higher amounts.

Keywords: carotenoids, fruits, Rosaceae

INTRODUCTION

Rosaceae, the rose family of flowering plants (order Rosales), is primarily found in the north temperate zone and occurs in a wide variety of habitats, members beeing generally woody plants, mostly shrubs or small to medium-size trees, some of which are armed with thorns, spines, or prickles. The Rosaceae family provides most of the world's well-known temperate fruit crops classified as pome and stone fruits according to their fruit morphology. With the global tendency of consumers' purchasing preferences to shift from large-scale commodity fruit crops (e.g., apples, citrus, and pears) to smaller unique fruit crops with increased nutritional value and a pleasing flavor, the rapid development of stone fruit crops (e.g., apricots, cherries, peaches and plums) has come to the forefront. (Folta and Gardiner, 2009).

Carotenoids, a class of isoprenoid compounds, are one of the most important active principles in Rosaceae family fruits, natural pigments that provide yellow to orange and red hues to plants, indispensable for plant life. They play essential roles in photosynthesis, and they serve as precursors to the plant hormones abscisic acid (ABA) and strigolactones. In addition, these compounds furnish many flowers and fruit with attractive yellow, orange and red colors. The compositions of carotenoids in flowers vary widely among plant species and cultivars. Pale to deep yellow flowers of most plants mainly accumulate xanthophylls (Zhu et al., 2010; Ohmiya, 2011).

Carotenoids are colored terpenes synthesized in plants, algae and some yeasts and bacteria. In plants and algae, these lipophilic molecules exert functional roles in hormone synthesis, photosynthesis, photomorphogenesis and photoprotection and are synthesized in the plastids, such as chloroplasts and chromoplasts (Li and Yuan, 2013).

Plant chloroplasts have a remarkably similar carotenoid composition in photosynthetic tissues, with lutein (45% of the total), β-carotene (25–30%), violaxanthin (10–15%), and neoxanthin (10–15%) as the most abundant carotenoids. In chloroplasts, carotenoids as well as other pigments, such as chlorophyll a and b, are localized and accumulate in the functional pigment-binding protein structures embedded in photosynthetic (thylakoid) membranes (Cunningham and Gantt, 1998), specifically near the reaction center of photosystem II in the light harvesting complexes (LHC). Carotenoids act as accessory pigments in the LHC, where they absorb light in a broader range of the blue spectrum (400-500 nm) than chlorophyll, and they transfer the absorbed energy to chlorophyll a during photosynthesis (Britton, 1995; Schmid, 2008).

The carotenoids as above mentioned are isoprenoid compounds, biosynthesized by tail-to-tail linkage of two C20 as seen in Figure 1, forming a complex of 40-carbon skeleton of isoprene units (Figure 2).

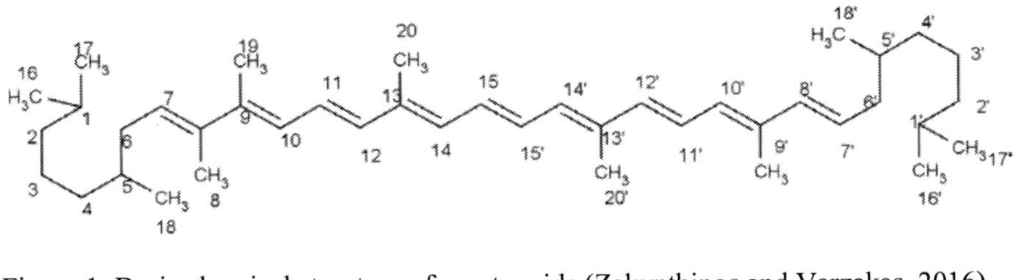

Figure 1. Basic chemical structure of carotenoids (Zakynthinos and Varzakas, 2016).

The structure may be cyclized at one or both ends, may have various hydrogenation levels, or may possess oxygen-containing functional groups. Lycopene and carotene are examples of acyclized and cyclized carotenoids, respectively. Carotene, β-carotene, and-cryptoxanthin are able to function as provitamin A. Each enzymatic step from phytoene to lycopene adds one double bound to the molecule, resulting in lycopene, which is a symmetrical molecule containing 13 double bonds. The biosynthetic step after lycopene involves enzymatic cyclization of the end groups, which results in γ-carotene (one β- ring) and β-carotene (two β-rings). The addition of oxygen to the molecule leads to the formation of xanthophylls (Zakynthinos and Varzakas, 2016).

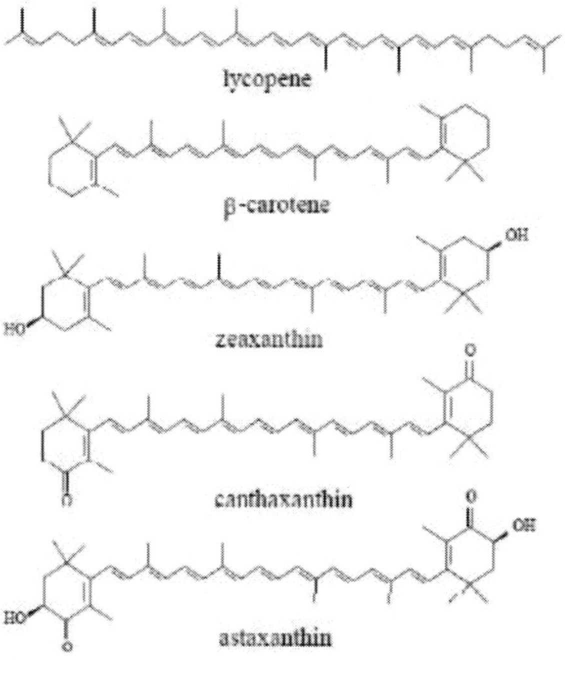

Figure 2. Carotenoids, molecular structures of 40-carbon skeleton of isoprene units, important for the antioxidant capacity of vegetal products (Zakynthinos and Varzakas 2016).

Additionally, they possess antioxidant properties and act as scavengers of reactive oxygen species from photooxidative damage by quenching singlet oxygen produced from the chlorophyll triplet in the reaction center of photosystem II (Takano et al., 2005; Telfer, 2005).

This antioxidant characteristic is a result of the conjugate bonds of the polyene chain, which permit the absorption of excess energy from other molecules (Britton, 1995; Britton et al., 1998; Nelson et al., 2003).

Birds, fish and crustaceans utilize carotenoids for pigmentation and nutrition. For example, the cetocarotenoid astaxanthin is responsible for the orange color of salmon meat and lobster shells (reviewed in Grotewold, 2006). Carotenoids serve also as pigments in several ornamental plants, in the cosmetic and food industries (Klaüi and Bauernfeind, 1981) and are employed as poultry and fish feed additives (reviewed in Bjerkeng, 2000).

Carotenoids are important in human health since they serve as pro-vitamin A and are antioxidants with protective activity against various illnesses. Also, carotenoids are precursors of carotenoid derived molecules termed apocarotenoids, which include isoprenoids with important functions in plant environment interactions such as the attraction of pollinators and the defense against pathogens and herbivores.

Apocarotenoids, also include volatile aromatic compounds that act as repellents, chemoattractants, growth simulators and inhibitors, as well as the phytohormones abscisic acid and strigolactones.

Carotenoid biosynthesis and identification of steps that control the flux through the pathway, the rate of biosynthesis, sequestration and the storage capacity of the cell, along with the rate of carotenoid catabolism, all play a significant role in determining the levels of carotenoid accumulation in plant tissues and organs. In Figure 3 a compressed version of the carotenoid biosynthesis pathway is presented.

The bioavailability of carotenoids in foods and in commercial preparations varies widely. Only about 5% of the carotenoids in whole, raw vegetables, for example, is absorbed by the intestine, whereas 50% or more of the carotenoids is absorbed from micellar solutions. Thus, the physical form in which the carotenoids is presented to intestinal mucosal cells is of crucial importance. In this way, the release of carotenoids from food matrix, their dispersion within the digestive tract, and their solubilization in mixed micelles are important steps for carotenoids bioaccessibility.

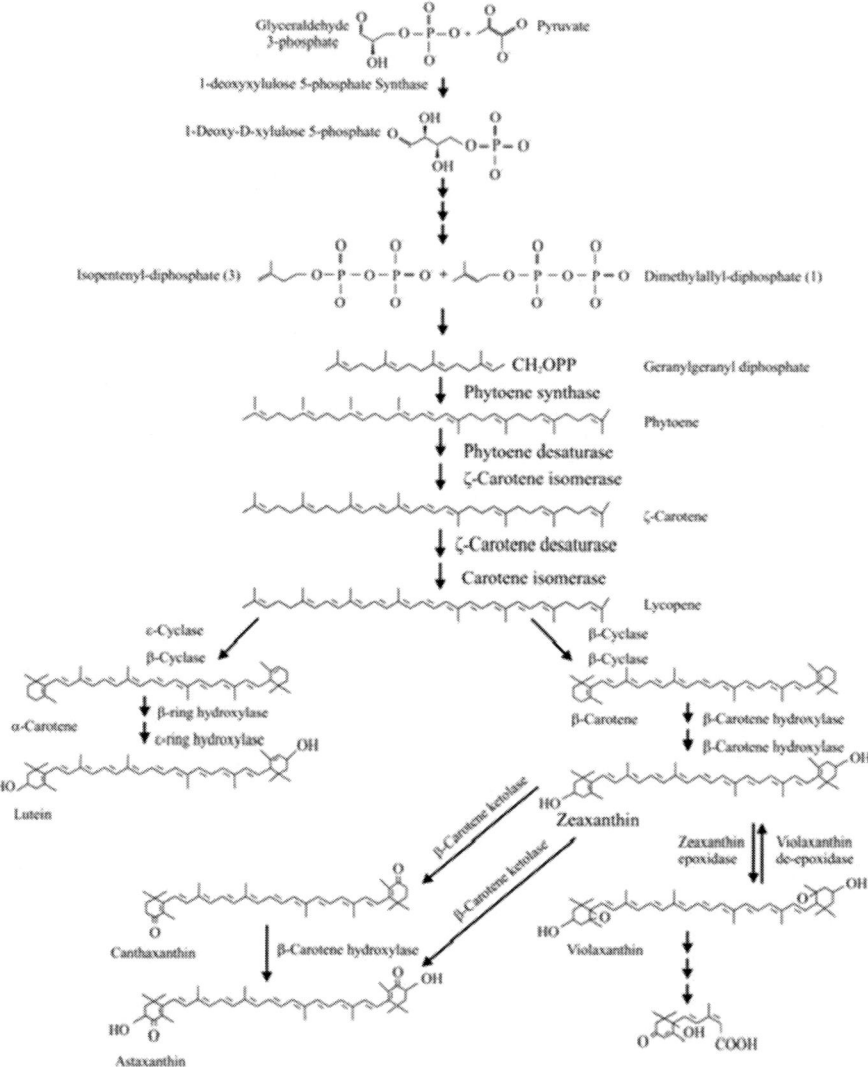

Figure 3. Carotenoid biosynthesis pathway
(Hannoufa and Hossain, AOCS Lipid Library).

Rosaceae Fruits as a Valuable Source of Carotenoids 117

ROSACEAE FAMILY: AN OVERVIEW

Rosaceae Juss. is the 19th largest plant family, with an estimated 95 - 100 genera, consisting of 2,830 - 3,100 species. The exact phyllogenetic relationships within the family are still subjected to dispute, however, the three recognized subfamilies are Dryadoideae, Rosoidaeae (roses and relatives) and Amygdaloideae/Spiraeoideae/Maloideae (sometimes separate subfamilies). The latter group consists of numerous tribes, hosting many common fruit trees and shrubs. Rosaceae fruits are variable, such as, drupe (*Prunus*), collection of achenes (*Potentilla*), berry (*Eriobotrya japonica*), pome (*Pyrus*) (Hummer and Jannick, 2009).

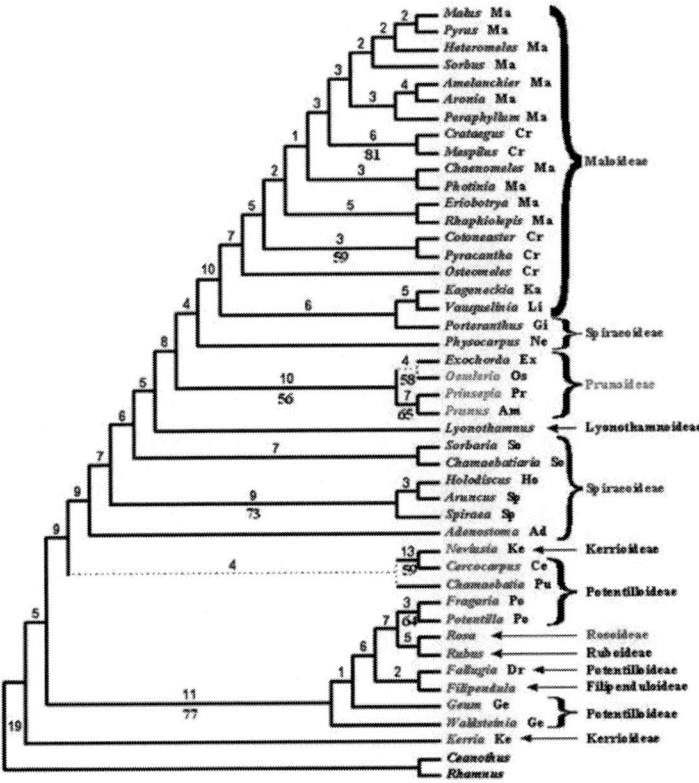

Figure 4. Proposed simplified phyllogenetic tree of Rosaceae family, with the main genera (Hummer and Jannick, 2009).

CAROTENOIDS IN ROSACEAE FRUITS

Rosehips

There are over 100 species in the genus *Rosa*, widely distributed throughout the Northern Hemisphere. With different appearances, many are cultivated as ornamental plants, for their flowers. However, their fruits, known as hips (pseudo-polyachenes), are edible and few species are grown for culinary purposes.

The most common is *Rosa canina* (dog rose), commonly growing as a wild or cultivated shrub in many countries, with 1.5-2 cm hips. They are consumed raw, as juice, syrup, marmalade, jelly, teas, etc. Dog rose hips are also used in folk medicine, being known for their anti-inflammatory, anti-rheumatic, antipyretic, laxative and diuretic properties. They are also useful against colds, influenza, ulcer, billiary and kidney stones (Ghazghazi et al., 2010). Antimutagenic and anticarcinogenic properties were also reported (Andersson, 2009).

A quantitative assay on *Rosa canina* fruits (Romanian cultivars) found percentages of 20.8% β-carotene, 27.8% lycopene, 23.6% rubixanthin, 11.3% lutein and zeaxanthin, 5.5% β-criptoxanthin (Hodișan et al., 1997). However, data show a high variation of carotenoid profiles among regional varieties: a study on two Tunisian varieties found a 5.36-7 β-carotene: lycopene ratio (Ghazghazi et al., 2010). An extensive study on *Rosa canina* varieties in Northern Iran found a preeminence of lycopene in most analyzed fruits, with lycopene: β-carotene ratios ranging from 1.04 to 7.65 (Shameh et al., 2019).

Rosa rugosa (Japanese rose), with its 2-3 cm, tomato-like fruits, is the second most important edible rose in Eurasia. Another important species is *Rosa rubiginosa* (syn. *Rosa eglanteria*) – sweet briar – an Eurasian species, used for herbal teas in some European countries and juices and jams in Turkey (Al-Yafeai, 2019).

Other common species include *Rosa pimpinellifolia* (syn. *Rosa spinosissima* – scotch rose), *Rosa dumalis* (glaucous dog rose). Most species contain, as their main fruit carotenoids, lutein, zeaxanthin and

β-carotene. In addition to those, *Rosa pimpinellifolia* was found to contain amounts of neochrome, neoxanthin and violaxanthin, while *Rosa rubiginosa* and *Rosa dumalis* contain rubixanthin, lycopene, prolycopene, γ-and ζ-carotene. Ripening stage significantly influences the total amount and the distribution of various carotenoid compounds in rosehips (Andersson, 2009). *Rosa villosa* (apple rose) fruits also contain high levels of carotenoids (see Table 1).

Table 1. Average of total carotenoids content in fruit pulp of various rose species

Species	Total carotenoids content (mg/kg)	Reference
Rosa canina	78.5 (DW)	Hodişan et al., 1997
	>24.8-31.8 (DW)	Ghazghazi et al., 2010
	4.67-20.05 (FW)	Shameh et al., 2019
	86.7 (DW)	Medveckienė et al., 2020
Rosa rugosa	165.8-252.9 (DW)	Medveckienė et al., 2020
Rosa rubiginosa	726.00 (DW)	Andersson, 2009
Rosa pimpinellifolia	297.1 (DW)	Andersson, 2009
Rosa dumalis	629.3 (DW)	Andersson, 2009
Rosa villosa	495.1 (DW)	Medveckienė et al., 2020
Rosa moschata	3.03-11.20 (FW)	Shameh et al., 2019
Rosa webbiana	6.51-7.07 (FW)	Shameh et al., 2019
Rosa damascena	20.01 (FW)	Shameh et al., 2019
Rosa hemisphaerica	9.36 (FW)	Shameh et al., 2019

Apples, Pears, Quinces

One of the largest taxonomic groups in the Rosaceae family is the Maleae tribe, consisting of woody plants, usually bearing pomes – a type of accessory, polycarpellary fruits. Among them, some species are characterized by their large pmes, these being some of the most commonly consumed fruits in temperate regions of the world: apples, pears, quinces and their relatives.

While undoubtedly a source of valuable nutrients, apples (*Malus domestica*) are not particularly rich in carotenoid compounds. There is a significant difference between total carotenoid contents in the brightly-colored fruit peel (58.72 - 1510.77 mg/kg DW, depending on cultivar) and fruit flesh (only 14.80 - 71.57 mg/kg DW; Delgado-Pelayo et al., 2014).

Table 2. Average of total carotenoids content in fruit pulp of major pome-bearing Rosaceae

Species	Total carotenoids content (mg/kg)	Reference
Malus domestica	14.80-71.57 (DW)	Delgado-Pelayo et al., 2014
	6.54-10.90 (FW)	Khandaker et al., 2013
	<1 (FW)	Perry et al., 2009
	5-55 (DW)	Mureşan et al., 2017
Malus sylvestris	>9.2 (DW)	Aziz et al., 2013
Pyrus communis	<1 (FW)	Olmedilla et al., 1998
	30-150 (DW)	Kolniak-Ostek et al., 2020
Cydonia oblonga	34.2 (FW)	Ponder and Hallman, 2017
Chaenomeles japonica	18.7-37.4 (FW)	Ponder and Hallman, 2017
Chaenomeles × superba	59.7 (FW)	Ponder and Hallman, 2017
Crataegus pentagyna	32.7 (DW)	Abbasova and Novruzov, 2013
Crataegus eriantha	29.8 (DW)	Abbasova and Novruzov, 2013
Crataegus caucasica	36.8 (DW)	Abbasova and Novruzov, 2013
Crataegus curvisepala	22.7 (DW)	Abbasova and Novruzov, 2013
Crataegus sulfurea	92.67-98.40 (FW)	Pérez-Lainez et al., 2019
Crataegus tracyi	11.84-38.68 (FW)	Pérez-Lainez et al., 2019
Crataegus aurescens	13.77-42.35 (FW)	Pérez-Lainez et al., 2019
Crataegus baroussana	5.71-38.10 (FW)	Pérez-Lainez et al., 2019
Crataegus rosei	22.26 (FW)	Pérez-Lainez et al., 2019
Crataegus gracillior	14.41-16.92 (FW)	Pérez-Lainez et al., 2019
Crataegus greggiana	10.71 (FW)	Pérez-Lainez et al., 2019
Crataegus mexicana	13.53-33.61 (FW)	Pérez-Lainez et al., 2019
Crataegus cuprina	38.71 (FW)	Pérez-Lainez et al., 2019
Eriobotrya japonica	16.40 (FW)	Olmedilla et al., 1998
	0.93-21.57 (FW)	Hasegawa et a., 2010

An extensive study on multiple commercial varieties found that green-skinned cultivars tend to have higher amount of carotenoids in their pulp, compared to yellow and, especially, red-skinned apples. Xantophyll esters (mostly diesters) are the dominant carotenoid pigments in fruit flesh (39-95%), with higher concentrations in red and yellow-skinned apples.

Apple peel, on the other hand, is dominated by lutein, violaxanthin, neoxanthin and β-carotene (in some red and yellow varieties, violaxanthin and neoxanthin are the main carotenoid pigments). The difference is due to different storage organelles: fruit peel contains mostly typical chloroplasts (also containing high amonts of chlorophylls), while chromoplasts dominate in the flesh (Delgado-Pelayo et al., 2014).

As with any other fruit, carotenoid content is not constant; it depends on ripening stage. A study conducted on apple peels of different varieties showed that, after reaching a maximum, both chlorophylls and carotenoids follow a steep decrease during on-tree ripening process (August-September). However, in case of detached fruits, after an initial lag phase, carotenoids increased again by 40% (Solovchenko et al., 2005).

The European crab apple, *Malus sylvestris* has a similar carotenoid content and profile to that of the cultivated species (Aziz et al., 2013).

Another important fruit is pear, with its main species, *Pyrus communis*. Like apples, pear pulp tends to be extremely poor in carotenoids, lutein being dominant in some cultivars, β-carotene in others (Olmedilla et al., 1998, Kolniak-Ostek et al., 2020).

Quinces (*Cydonia oblonga*) and related Japanese quinces (*Chaenomeles* sp. – though cultivated mostly for ornamental purposes, but also beverages and marmalades) also contain small amounts of carotenoids, especially zeaxanthin, followed by lutein and β-carotene (Ponder and Hallman, 2017).

Apart from these main species, there are some other Rosaceae bearing relatively large pomes, of lower alimentary importance. Among them, the around 150 species of hawthorns (*Crataegus* sp.). While the common Eurasian hawthorn (*Crataegus monogyna*) is usually cultivated for ornamental and some folk medicine uses, the fruits of some Mexican species (collectively known as "tejocote") are oftenly consumed raw, cooked, as jams, juices or in alcoholic beverages (Pérez-Lainez et al., 2019).

Hawthorn fruits have total carotenoid amounts comparable to those in other pomes mentioned above. β-carotene, followed by zeaxanthin and α-carotene are the main compounds (Abbasova and Novruzov, 2013).

Finally, loquat, or Japanese/Chinese plum (*Eriobotrya japonica*), with its large, pectin-rich pome, is a popular fruit in warmer areas of the Globe, used for jams, jellies, chutneys, juices, smoothies, eaten raw or cooked. β-

carotene is the dominant carotenoid pigment, followed by β-cryptoxanthin (Hasegawa et al., 2010).

Other Pomes: Chokeberries, Rowans, Rocksprays, Firethorns

Apart from the species mentioned above, the Maleae tribe comprises several genera of trees and shrubs producing small-sized, berry-like pomes, with alimentary and medicinal uses in various parts of the world.

Black chokeberries (*Aronia melanocarpa*), natives of Northeastern America, are becoming more and more popular throughout the world. These small pomes are used for syrups, spreads, jellies, herbal teas or as a coloring agent. However, while chokeberries are rich in anthocyanins, their total carotenoid content is moderate, with β-carotene, violaxanthin and β-cryptoxanthin being the dominant ones (Kulling and Rawel, 2008).

Even more popular and widely cultivated are rowanberries (*Sorbus* sp.). Their astringent, Vitamin C-rich fruits are commonly used for jellies, juices, chutneys, liqueur or teas. Taxonomy of cultivated rowans is rather intricate. There are numerous cultivars grown throughout the world and, while most derive from the common Eurasian rowan or mountain-ash (*Sorbus aucuparia*), many are complex hybrids, involving ancestors like *Sorbus domestica*, *Sorbus aria*, *Sorbus cashmiriana*, *Sorbus commixta* or *Aronia arbutifolia*. Hence the extreme variety of their biochemical composition and the extreme range of total carotenoid content (Zymonė et al., 2018). Service trees, *Sorbus torminalis* are also consumed in some parts of the world, raw or in jams, syrups or used for flavouring alcoholic beverages.

Major differences regarding the proportion of various compounds are found even between populations of the same species. Some studies point to a dominance of β-carotene and β-cryptoxanthin in both mountain-ash (*Sorbus aucuparia*) and whitebeam (*Sorbus aria*) fruits (O'Sullivan et al., 2011). However, lycopene and β-carotene seem to be the dominant carotenoid pigments in other *Sorbus aria* varieties, with some variable amounts of β-cryptoxanthin, lutein and α-carotene in whitebeam (Petkova et al., 2020). Also, data from some wild *Sorbus aucuparia* populations show an entirely different profile, with large amounts of lutein epoxide (28%), capsanthin (24.5%), lycoxanthin (19%), capsorubin, isozeaxanthin, β-cryptoxanthin and only 4.5% β-carotene (Czeczuga, 1978).

Table 3. Average of total carotenoids content in fruit pulp of minor pome-bearing Rosaceae

Species	Total carotenoids content (mg/kg)	Reference
Aronia melanocarpa	48.6 (FW)	Kulling and Rawel, 2008
Sorbus commercial hybrids	39-2,659 (DW)	Zymonė et al., 2018
	72.5-130.4 (DW)	Kampuss et al., 2009
Sorbus aucuparia	120 (DW)	Czeczuga, 1978
	95.2 (DW)	Kampuss et al., 2009
Sorbus aria	37.17 (DW)	Petkova et al., 2020
Sorbus torminalis	159 (DW)	Popoviciu and Negreanu-Pîrjol, 2019
Cotoneaster lacteus	23.81-75.03 (DW)	Popoviciu et al., 2019
Cotoneaster salicifolius	271.31-346.04 (DW)	Popoviciu et al., 2020a
Cotoneaster horizontalis	315.47-433.62 (DW)	Popoviciu et al., 2020b
Cotoneaster microphyllus	165.57-189.83 (DW)	Popoviciu et al., 2020b
Pyracantha angustifolia	>63 (DW)	Zechmeister and Schroeder, 1942
Pyracantha coccinea	278-545 (DW)	Popoviciu et al., 2020c
Pyracantha crenulata	>22 (DW)	Pal et al., 2013

Other related species feature small-sized berries. *Cotoneaster* species have little culinary significance, mostly because ther seeds are rich in mildly toxic cyanogenic glycosides and difficult to remove. However, some species, such as *Cotoneaster microphyllus* (rockspray) and *Cotoneaster multiflorus* are used in East Asian folk medicine, against skin irritation, infections, gastritis, bronchitis, biliar disfunction, vascular diseases or irregular menstruation. Others are cultivated solely for ornamental purposes. Studies have shown that they sometimes contain significant levels of total carotenoids (Popoviciu et al., 2019).

Firethorns (*Pyracantha* sp.) are decorative shrubs whose fruits can also have culinary uses (marmalade, jams, jellies, sauces, especially *Pyracantha coccinea*) or medicinal properties (tonic, cardiotonic, diuretic, especially *Pyracantha crenulata*) are also a valuable source of carotenoids (Popoviciu et al., 2020c). While in *Pyracantha crenulata* pomes lycopene and β-carotene dominate (Pal et al., 2013), in *Pyracantha angustifolia* prolycopene and pro-γ-carotene form the bulk of the total carotenoid content (Zechmeister and Schroeder, 1942).

Stone Fruits

One of the largest and most widely cultivated Rosaceae are those in the Amygdaleae tribe, namely the *Prunus* genus, with numerous species producing drupes (stone fruits): plums, peaches, apricots, cherries, almonds and their relatives.

Usually, the fruits of the European plum tree (*Prunus domestica*) are considered to possess one of the lowest total carotenoid amounts. The exact amount variates among cultivars. β-carotene, β-cryptoxanthin, lutein, α-carotene and lycopene may have comparable shares of the total carotenoid amount in some varieties (Leong and Oey, 2012), while in others β-carotene and lutein form the bulk (with either one dominant Olmedilla et al., 1998; Yoon et al., 2012). Finally, there are cultivars whose fruits contain mostly β-carotene, followed by β-cryptoxanthin. There are also major differences between plum skin and flesh, with 2-7 times higher amounts in the skin (a feature common to all stone fruits; Gil et al., 2002).

A similar and widely cultivated species is the Japanese/Chinese plum (*Prunus salicina*), a native of China, now commonly grown in East Asia, Australia and North America. It also features low amounts of carotenoids, with variate composition, β-carotene, lutein, β-cryptoxanthin and zeaxanthin being dominant (González-Flores et al., 2011, Bobrich et al., 2014).

Among other related species are *Prunus cocomilia* (Italian plum), *Prunus cerasifera* (cherry plum), and *Prunus spinosa* (blackthorn, sloe), both with a carotenoid profile dominated by β-carotene (O'Sullivan et al., 2011; Gironés-Vilaplana et al., 2014; Kaulmann et al., 2014).

Table 4. Average of total carotenoids content in fruit pulp of drupe-bearing Rosaceae

Species	Total carotenoids content (mg/kg)	Reference
Prunus domestica	2 (FW)	Olmedilla et al., 1998
	>0.45-4.39 (DW)	Gil et al., 2002
	170 (DW)	Leong and Oey, 2012
	1.7 (FW)	Yoon et al., 2012
	2.47-19.45 (DW)	Kaulmann et al., 2014
	2.31 (FW)	Vinholes et al., 2016
Prunus cocomilia	19.03 (FW)	Kaulmann et al., 2014
Prunus salicina	10.46 (DW)	González-Flores et al., 2011
	0.8-1 (FW)	Bobrich et al., 2014
Prunus cerasifera	19.56 (DW)	Kaulmann et al., 2014
Prunus spinosa	>10 (FW)	O'Sullivan et al., 2011
	>8 (FW)	Gironés-Vilaplana et al., 2014
Prunus persica	1.88 (FW)	Olmedilla et al., 1998
	0.04-1.68 (DW)	Gil et al., 2002
	140 (DW)	Leong and Oey, 2012
	49.1-145.50 (DW)	Brown et al., 2014
	<11.60 (FW)	Vinholes et al., 2016
Prunus persica var. *nectarina*	0.04-1.46 (DW)	Gil et al., 2002
	430 (DW)	Leong and Oey, 2012
	1.62 (FW)	Vinholes et al., 2016
	0.9 (FW)	Yoon et al., 2012
Prunus persica var. *platycarpa*	619.00 (FW)	Loizzo et al., 2015
Prunus armeniaca	1.76 (FW)	Olmedilla et al., 1998
	16.00-19.70 (FW)	El-Badawy and El-Salhy, 2011
	970 (DW)	Leong and Oey, 2012
	196.40-893.30 (FW)	Vinha et al., 2012
	5.60 (DW)	Čanadanović-Brunet, 2013
	15.12-16.50 (FW)	Vinholes et al., 2016
Prunus mume	0.7 (FW)	Yoon et al., 2012
Prunus avium	0.78 (FW)	Olmedilla et al., 1998
	90 (DW)	Leong and Oey, 2012
	0.011 (FW)	Vinholes et al., 2016
Prunus cerasus	0.60 (FW)	Yoon et al., 2012
	1.70-3.90 (DW)	Radenkovs and Feldmane, 2017

Peaches (*Prunus persica*) come in many different varieties: peaches, nectarines (*Prunus persica* var. *nucipersica/nectarina*), flat peaches (*Prunus persica* var. *platycarpa*), with extremely diverse cultivars and hybrids. Thus, their carotenoid contents and profile is extremely diverse.

For instance, common peaches can contain 7 times more carotenoid compounds than nectarines, being coles to the known extremes for this plant genus (Vinholes et al., 2016; this is not the case for all nectarine cultivars, see Leong and Oey, 2012).

Even among the same category, cultivars can be radically different: while violaxanhtin is the dominant compound in some, others feature a higher amount of zeaxanthin, lutein, lutein epoxide or others. Zeaxanthin, β-carotene and β-cryptoxanthin are also present (Brown et al., 2014). Generally, white flesh peaches and nectarines have much lower carotenoid contents than yellow-red flesh ones. All peaches also show major differences between fruit skin and flesh, with the former having 2-14 times higher amounts (Gil et al., 2002).

Other popular stone fruits are apricots (*Prunus armeniaca*) and related Japanese apricot (*Prunus mume*). Apricots are known to have one of the richest (although extremely variable) carotenoid inventory among Amygdalae (Vinholes et al., 2016), β-carotene, β-cryptoxanthin, α-carotene, lycopene and lutein beeing the main ones (Leong and Oey, 2012).

Nutritional highly appreciated, cherry species are native to the Northern Hemisphere, where any of various trees belonging to the genus Prunus are widely grown, the commercial production including three types of cherries mainly grown for their fruits, sweet cherries (*Prunus avium*), sour cherries (*Prunus cerasus*), and grown to a much smaller extent, the dukes, which are crosses of sweet and sour cherries.

Sweet cherry (*Prunus avium*, L.) and sour cherry (*Prunus cerasus*, L.) trees are two of the economically most valuable stone fruit tree species. The species *Prunus avium* (syn. *Cerasus avium*) Moench. with the following varieties: *sylvestris* Ser., wild cherry, 1 cm in diameter, black, slightly juicy, sweet-bitter and early ripening; *juliana* (L.) Pojarkova, with medium fruits, soft pulp, early and medium ripening; *duracina* (L.) Pojarkova, stony cherries, large fruits, hard, stony pulp, slightly juicy and late ripening, belongs to the Rosaceae family, are considered valuable nutritional products for humans (Negreanu–Pirjol B. et al., 2009, 2014). With rich content in vitamins, minerals, polyphenols, carotenoids and many other compounds, *Prunus avium* (syn. *Cerasus avium*) Moench. fruits extracts, may be useful as a valuable antioxidant, an adjuvant in the treatment or in preventing many diseases which arise from oxidative stress action of free radicals (Negreanu-Pirjol et al., 2021; Negreanu-Pirjol T. et al., 2014, 2015).

The regular consumption of cherries is related to a balanced diet and broad spectrum of health benefits due to the great quantity of bioactive or nutraceutical compounds present in this fruit (González-Gómez et al., 2010; Tomás-Barberán et al., 2013). By late fourteenth and early fifteenth centuries, cherries were included in an Italian medical book, because they were considered to be fruit with health benefits and were probably collected for medical purposes (Janick, 2011; Serradilla et al., 2016).

Sweet fruit is a fleshy drupe (stone fruit) about 2 cm in diameter, and varies in colour from yellow through red to nearly black. The acid content of the sweet cherry is low but the higher acid content of the sour cherry produces its characteristic tart flavour. Sour cherry fruit is round to oblate in shape, is generally dark red in colour, and has so much acid that it is not appealing for eating fresh. The duke cherries are intermediate in both tree and fruit characteristics. Usually, yellow skin cultivars fruits contain the highest concentrations of carotenoids, beeing strong positive correlations between the identified bioactive compounds and fruits antioxidant activity (Srednicka-Tober et al., 2019).

The fruits of all varieties provide carotenoids, although different cultivars of sweet cherries emphasize a high variability in total carotenoids profiles, between 5.90 - 9.02 g 100 g^{-1} FW, respectively lutein 0.24 - 2.73 mg 100 g^{-1} FW, zeaxanthin 0.11 – 2.12 mg 100 g^{-1} FW, α-carotene 1.56 – 3.05 mg 100 g^{-1} FW, β-carotene 1.66 – 2.88 mg 100 g^{-1} FW. The authors concluded that sour cherries are generally richer in phenolic compounds, especially anthocyanins and hydroxycinnamic acids and compared with sweet cherries with a deep purple color (rich in anthocyanins) showed high antioxidant activity, which confirms the antioxidant potential of these bioactive compounds. However, fruits cultivars with yellow skin and flesh (without or with low concentrations of anthocyanins) also showed significant antioxidant potential, correlated with the concentrations of carotenoids - a strong antioxidants group (Srednicka-Tober et al., 2019).

Sweet cherries and sour cherries are considered more poor sources of carotenoid pigments (Vinholes et al., 2016). Lutein is sometimes dominant, while in other cultivars, it has a similar share with $β$-carotene, $β$-cryptoxanthin, $α$-carotene and lycopene (Olmedilla et al., 1998; Leong and Oey, 2012), and they all should be, similarly to other types of cherries, recognized as a rich source of bioactive compounds with an antioxidant potential, thus human health-promoting and disease-preventive potential.

Other Rosaceae

The Rosaceae family comprises several other plants, with culinary or medicinal uses, having aggregate fruits, such as polydrupes and (pseudo-) polyachenes.

One of the largest and most variate genera is *Rubus*, including raspberries, blackberries, cloudberries, dewberries etc. Their fruits are fleshy polydrupes, formed of a variable number of drupelets (small stone fruits).

Rubus idaeus (European red raspberry) is considered a source of carotenoids, especially lutein and $β$-carotene, with minor amounts of neoxanthin, violaxanthin, and $α$-carotene (Abdul and Majeed, 2012). *Rubus occidentalis* (black raspberry) fruits, on the other hand, are dominated by carotenoid esters (Perkins-Veazie et al., 2020).

Blackberries are now considered to be part of several species, mostly cultivated being *Rubus plicatus* (syn. *Rubus fruticosus*; European blackberry). Blackberries can contain significant amounts of carotenoid compounds, especially $β$-carotene, lutein and zeaxanthin (Zia-ul-Haq et al., 2014). Its main South American relative is *Rubus sellowii*, cultivated throughout the continent and rich in lycopene (Teixeira et al., 2018).

Other species of culinary use include Andean raspberry (*Rubus glaucus*) (Mayorga and Grandes, 2014) and Chinese raspberry (*Rubus chingii*), the latter being dominated by apocarotenoids like β-citraurin and derivatives, plus lutein and zeaxanthin (Li et al., 2021).

Table 5. Average of total carotenoids content in fruit pulp of other Rosaceae

Species	Total carotenoids content (mg/kg)	Reference
Rubus idaeus	52.60 (DW)	Abdul and Majeed, 2012
	8-20 (DW)	Bradish et al., 2015
Rubus occidentalis	2.33 (DW)	Perkins-Veazie et al., 2020
Rubus plicatus	2.46 (FW)	Zia-ul-Haq et al., 2014
	122.78-222.15 (DW)	de Souza et al., 2018
Rubus sellowii	56 (DW)	Teixeira et al., 2018
Rubus glaucus	6.60 (FW)	Mayorga and Grandes, 2014
Rubus chingii	1,842-3,054 (FW)	Li et al., 2021
Rubus chamaemorus	540 (DW)	Czeczuga, 1978
	28.40 (DW)	Lashmanova et al., 2012
Rubus alpestris	21,860 (DW)	Abu Bakar et al., 2016
Rubus fraxinifolius	10,490 (DW)	Abu Bakar et al., 2016
Rubus moluccanus	9,690 (DW)	Abu Bakar et al., 2016
Fragaria × ananassa	0.18 (FW)	Olmedilla et al., 1998
	>0.05 (FW)	Khoo et al., 2011
Fragaria moschata	<0.10 (FW)	Yoon et al., 2012
Fragaria vesca	<0.04 (FW)	Pritwani and Mathur, 2017

Less cultivated are cloudberries (*Rubus chamaemorus*), popular in Scandinavian cuisine, rich in carotenoids, mostly β-cryptoxanthin, zeaxanthin, flavoxanthin, antheraxanthin and β-carotene, with variations among different plant populations (Czeczuga, 1978; Lashmanova et al., 2012).

Other *Rubus* species are used only in folk medicine, in various parts of the world, for illnesses like infertility, impotence, backache, poor eyesight, frequent urination, or as laxatives (Abu Bakar et al., 2016).

Fragaria genus contains several wild and cultivated species, with polyachenes featuring a large, fleshy, central receptacle. The common garden strawberry (*Fragaria×ananassa*) is a hybrid of *Fragaria virginiana* and *Fragaria chiloensis*, is nowadays the most popular member of the genus. Tissues in its receptacle contains a low amount of carotenoids, with a dominance of α-carotene, or lutein and β-carotene, depending on cultivar

(Olmedilla et al., 1998; Khoo et al., 2011). Very low amounts of carotenoids are also found in *Fragaria moschata* (musk strawberry) and and *Fragaria vesca* (wild/Alpine strawberry; Yoon et al., 2012; Pritwani and Mathur, 2017).

CONCLUSION

Carotenoids are the phytonutrients that impart a distinctive yellow, orange, and red color to various fruits and vegetables. The physico-chemical properties and the biological activities of carotenoids are intimately related to their structures. Amongst several dozen of antioxidants in the foods that we eat, most of these carotenoids have antioxidant activity. Carotenoids have been studied for their ability to prevent chronic disease, since the free radical theory of aging in chronic disease etiology remains pre-eminent.

β-carotene, lycopene, zeaxanthine and others carotenoids have antioxidant properties, but the antioxidant capability is variable depending on the *in vitro* system used. The antioxidant activity of these compounds can shift into a prooxidant effect, depending on such factors as oxygen tension or carotenoid concentration. Mixtures of carotenoids alone or in association with others antioxidants can increase their activity against lipid peroxidation. The isomeric form may be an important factor in the bioavailability and bioactivity of certain carotenoids but also the light on carotenogenic gene regulation in photosynthetic (leaves) and non-photosynthetic organs (flower and fruits) is important. Light acts as an inducer of photomorphogenesis and for carotenoid biosynthesis through photoreceptors and the activation of known transcription factors. Conventional studies focused on a specific gene or step in the carotenoid pathway combined with new technologies permitting an analysis of the entire pathway will be needed to understand the role of light upon carotenoid biosynthesis in diverse organisms.

Without doubt, aspects associated with the effect of light upon carotenoid biosynthesis regulation will be avenues warranting more intensive research efforts.

Many factors involved in carotenoid biosynthesis have been found by means of post-transcriptional gene silencing. Alternatively, the utilization of fruit-specific promoters (Corona et al., 1996), and light-positive and light-negative regulators (Davuluri et al., 2005) will allow the production of new varieties of plants with enhanced carotenoid accumulation specifically associated to fruits, without affecting the carotenoid biosynthesis regulation in non-target plant organs.

Also, optimizing carotenoids rich fruits quality attributes needs a better understanding of the interaction between genetics, environmental factors, and plant growth regulators to be able to gain full benefit of new preservation technologies.

The total antioxidant capacity registered in some fruits fluid extracts, sustain the possibility to use the bioactive compounds and the incorporation of bioactive compounds may provide advantages in food preservation and contributes to the development of functional foods. Application of bioactive compounds and encapsulation with edible coatings are promising techniques that also need further investigation, also with regard to diminishing off the flavors.

Carotenoids as natural pigments could be used by the industry as pharmaceuticals, nutraceuticals and animal feed additives, as well as colorants in cosmetics and special foods. In many food crops, especially the major staple crops contain only traces to low amounts of carotenoids. Through biotechnology significant progress has been made in developing food crops rich in carotenoids by altering the expression of carotenoids biosynthetic genes or using microorganisms to produce carotenoids by fermentation procedure. This strategy could be used to develop different kinds of fruits with improved nutritional value, helping consumers to incorporate vitamin A in their diets.

REFERENCES

Abbasova, T. Y. & Novruzov, E. N. (2013). Carotenoids from fruits of several *Crataegus* species. *Chemistry of Natural Compounds*, *49* (5), 965-966.

Abdul, D. A. & Majeed, S. N. (2012). Identification of antioxidant compounds in red raspberry (*Rubus idaeus*) fruit in Kurdistan region (north Iraq). *IOSR Journal of Applied Chemistry*, *2* (3), 6-10.

Abu Bakar, M. F., Ismail, N. A., Isha, A. & Ling, A. L. M. (2016). Phytochemical composition and biological activities of selected wild berries (*Rubus moluccanus* L., *R. fraxinifolius* Poir., and *R. alpestris* Blume). *Evidence-Based Complementary and Alternative Medicine*, *2016* (11), doi: 10.1155/2016/2482930.

Al-Yafeai, A. (2019). *In vitro investigations on biological activities and bioaccessibility of rosehip carotenoids*. Doctoral thesis, Friedrich Schiller University, Jena.

Andersson, S. C. (2009). *Carotenoids, tocochromanols and chlorophylls in sea buckthorn berries (Hippophae rhamnoides) and rose hips (Rosa sp.)*. Doctoral thesis, Swedish University of Agricultural Sciences, Alnarp.

Aziz, M., Anwar, M., Uddin, Z., Amanat, H., Ayub, H. & Jadoon, S. (2013). Nutrition comparison between genus of apple (*Malus sylvestris* and *Malus domestica*) to show which cultivar is best for the province of Balochistan. *Journal of Asian Scientific Research*, *3* (4), 417-424.

Benichou, M., et al. (2018). Postharvest technologies for shelf life enhancement of temperate fruits, Postharvest Biology and Technology of Temperate Fruits (eds Shabir Ahmad, M., Manzoor Ahmad, S. & Mohammad Maqbool, M.) 77–100 (*Springer*, 2018).

Bobrich, A., Fanning, K. J., Rychlik, M., Russell, D., Topp, B. & Netzel, M. (2014). Phytochemicals in Japanese plums: impact of maturity and bioaccessibility. *Food Research International*, *65*, 20-26.

Bradish, C. M., Yousef, G. G., Ma, G., Perkins-Veazie, P. & Fernandez, G. E. (2015). Anthocyanin, carotenoid, tocopherol, and ellagitannin content of red raspberry cultivars grown under field or high tunnel cultivation in the Southeastern United States. *Journal of the American Society for Horticultural Science, 140* (2), 163-171.

Brown, A. F., Yousef, G. C., Guzman, Y., Chebrolu, K. K., Werner, D. J., Parker, M., Gasic, K. & Perkins-Veazie, P. (2014). Variation of carotenoids and polyphenolics in peach and implications on breeding for modified phytochemical profiles. *Journal of the American Society for Horticultural Science, 139* (6), 676-686.

Cazzonelli, C. I. & Pogson, B. J. (2010). Source to sink: Regulation of carotenoid biosynthesis in plants. *Trends Plant Sci., 15,* 266-274 (DOI: 10.1016/j.tplants.2010.02.003).

Čanadanović-Brunet, J. M., Vulić, J. J., Ćetković, G. S., Djilas, S. M. & Tumbas Šaponjac, V. T. (2013). Bioactive compounds and antioxidant properties of dried apricot. *Acta Periodica Tehnologica, 44,* 193-205.

Czeczuga, B. (1978). The carotenoid content in certain plants from Abisko National Park (Swedish Lapland). *Acta Societatis Botanicorum Poloniae, 47* (3), 205-209.

Davuluri, G. R., van Tuinen, A., Fraser, P. D., Manfredonia, A., Newman, R., Burgess, D., Brummell, D. A., King, S. R., Palys, J., Uhlig, J., Bramley, P. M., Pennings, H. M. & Bowler, C. (2005). Fruit-specific RNAi-mediated suppression of DET1 enhances carotenoid and flavonoid content in tomatoes. *Nat Biotechnol., 23*(7), 890-5. doi: 10.1038/nbt1108. PMID: 15951803; PMCID: PMC3855302.

De Souza, A. V., Vieites, R. L., Gomes, E. P. & Vieira, M. R. S. (2018). Biochemical characterization of blackberry fruit (*Rubus* sp) and jellies. *Australian Journal of Crop Science, 12,* 624-630.

Delgado-Pelayo, R., Gallardo-Guerrero, L. & Hornero-Méndez, D. (2014). Chlorophyll and carotenoid pigments in the peel and flesh of commercial apple fruit varieties. *Food Research International, 65,* 272-281.

El-Badawy, H. E. M. & El-Salhy, F. T. A. (2011). Physical and chemical properties of canino apricot fruits during cold storage as influenced by some post-harvest treatments. *Australian Journal of Basic and Applied Sciences, 5* (9), 537-548.

Folta, K. M. & Gardiner, S. E. (2009). *Genetics and Genomics of Rosaceae* (Springer, 2009).

Ghazghazi, H., Miguel, M. G., Hasnaoui, B., Sebei, H., Ksontini, M., Figueiredo, A. C., Pedro, L. G. & Barroso, J. G. (2010). Phenols, essential oils and carotenoids of *Rosa canina* from Tunisia and their antioxidant activities. *African Journal of Biotechnology*, *9* (18), 2709-2716.

Gil, M. I., Tomás-Barberán, F. A., Hess-Pierce, B. & Kader, A. A. (2002). Antioxidant capacities, phenolic compounds, carotenoids, and vitamin C contents of nectarine, peach, and plum cultivars from California. *Journal of Agricultural and Food Chemistry*, *50* (17), 4976-4982.

Gironés-Vilaplana, A., Villaño, D., Baenas, N., Moreno, D. A. & García-Viguera, C. (2014). Blackthorn. In: Gironés-Vilaplana, A., Baenas, N., Villaño, D., Moreno, D.A. (eds.), *Iberian-American Fruits Rich in Bioactive Phytochemicals for Nutrition and Health*. Limencop, Alicante, Spain: 15-19.

González-Flores, D., Velardo, B., Garrido, M., González-Gómez, D., Lozano, M., Ayuso, M. C., Barriga, C., Paredes, S. D. & Rodríguez, A. B. (2011). Ingestion of Japanese plums (*Prunus salicina* Lindl. cv. Crimson Globe) increases the urinary 6-sulfatoxymelatonin and total antioxidant capacity levels in young, middle-aged and elderly humans: Nutritional and functional characterization of their content. *Journal of Food and Nutrition Research*, *50* (4), 229-236.

Hasegawa, P. H., de Faria, A. F., Mercadante, A. Z., Chagas, E. A., Pio, R., Lajolo, F. M., Cordenunsi, B. R. & Purgatto, E. (2010). Chemical composition of five loquat cultivars planted in Brazil. *Ciência e Tecnologia de Alimentos*, *30* (2), 552-559.

Hodişan, T., Socaciu, C., Ropan, I. & Neamţu, G. (1997). Carotenoid composition of *Rosa canina* fruits determined by thin-layer chromatography and high-performance liquid chromatography. *Journal of Pharmaceutical and Biomedical Analysis*, *16*, 521-528.

Hummer, K. E. & Jannick, J. (2009). Rosaceae: Taxonomy, Economic Importance, Genomics. In: Folta K. M., Gardiner S. E. (eds.), *Genetics* and Genomics of Rosaceae. *Plant Genetics and Genomics: Crops and Models*, vol 6. Springer, New York, USA: 1-17.

Kampuss K., Kampuse S., Berna E., Kruma Z., Krasnova I. & Drudze, I. (2009). Biochemical composition and antiradical activity of rowanberry (*Sorbus* L.) cultivars and hybrids with different Rosaceae L. cultivars. *Latvian Journal of Agronomy*, *12*, 59–65.

Kaulmann, A., Jonville, M. C., Schneider, Y. J., Hoffmann, L. & Bohn, T. (2014). Carotenoids, polyphenols and micronutrient profiles of *Brassica oleraceae* and plum varieties and their contribution to measures of total antioxidant capacity. *Food Chemistry*, *155*, 240-250.

Kearney, J., Food consumption trends and drivers. (2010). *Philos. Trans. R. Soc. B: Biol. Sci.*, *365*, 2793–2807.

Khandaker, M. M., Boyce, A. M., Osman, N., Golam, F., Rahman, M. M. & Sofian-Azirun, M. (2013). Fruit Development, Pigmentation and biochemical properties of wax apple as affected by localized application of GA3 under field conditions. *Brazilian Archives of Biology and Technology*, *56* (1), 11-20.

Khoo, H. E., Prasad, K. N., Kong, K. W., Jiang, Y. & Ismail, A. (2011). Carotenoids and their isomers: color pigments in fruits and vegetables. *Molecules*, *16*, 1710-1738

Kolniak-Ostek, J., Kłopotowska, D., Rutkowski, K. P., Skorupińska, A. & Kruczyńska, D. E. (2020). Bioactive compounds and health-promoting properties of pear (*Pyrus communis* L.) fruits. *Molecules*, *25* (4444), doi: 10.3390/molecules25194444.

Kulling, S. E. & Rawel, H. M. (2008). Chokeberry (*Aronia melanocarpa*) – A review on the characteristic components and potential health effects. *Planta Medica*, *74*, 1625-1634.

Lashmanova, K. A., Kuzivanova, O. A. & Dymova, O. V. (2012). Northern berries as a source of carotenoids. *Acta Biochimica Polonica*, *59* (1), 133-134.

Leong, S. Y. & Oey, I. (2012). Effects of processing on anthocyanins, carotenoids and vitamin C in summer fruits and vegetables. *Food Chemistry*, *133* (4), 1577-1587.

Li, X., Sun, J., Chen, Z., Jiang, J. & Jackson, A. (2021). Characterization of carotenoids and phenolics during fruit ripening of Chinese raspberry (*Rubus chingii* Hu). *RSC Advances*, *11*, 10804-10813.

Loizzo, M. R., Pacetti, D., Lucci, P., Núñez, O., Menichini, F., Frega, N. G. & Tundis, R. (2015). *Prunus persica* var. *platycarpa* (Tabacchiera peach): bioactive compounds and antioxidant activity of pulp, peel and seed ethanolic extracts. *Plant Foods for Human Nutrition*, *70*, 331-337.

Mayorga, S. E. & Grandes, B. B. (2014). Andean raspberry. In: Gironés-Vilaplana, A., Baenas, N., Villaño, D., Moreno, D. A. (eds.), *Iberian-American Fruits Rich in Bioactive Phytochemicals for Nutrition and Health.* Limencop, Alicante, Spain: 15-19.

Medveckienė, B., Kulaitienė, J., Vaitkevičienė, N. & Hallman, E. (2020). Carotenoids, polyphenols, and ascorbic acid in organic rosehips (*Rosa* spp.) cultivated in Lithuania. *Applied Sciences*, *10* (5337), doi: 10.3390/app10155337.

Muresan, E. A., Muste, S., Mureşan, C. C., Mudura, E., Păucean, A., Stan, L., Vlaic, R. A., Cerbu, C. G. & Muresan, V. (2017). Assessment of polyphenols, chlorophylls, and carotenoids during developmental phases of three apple varieties. *Romanian Biotechnological Letters*, *22* (3), 12546-12553.

Negreanu–Pirjol, B., Popescu, A., Bucur, L., Negreanu–Pirjol, T. & Arcus, M. (2009). Preliminary data regarding the study of some *Prunus serotina* Ehrh. fruits extract, *Rev. Med. Chir. Soc. Med. Nat. of Iassy*, *113*, (2), Supplement No. 4, pp. 282-286.

Negreanu-Pirjol, B. S., Negreanu–Pirjol, T., Bratu, M. M., Roncea, F., Miresan, H., Jurja, S., Paraschiv, G. M. & Popescu, A. (2014). Antioxidative activity of indigen bitter cherry fruits extract corellated with polyfenols and minerals content, 14th International Multidisciplinary Scientific GeoConferences "Surveying Geology & mining Ecology Management – SGEM 2014", Section: *Advances in Biotechnology*, Vol. *I*, pp. 239 – 244.

Negreanu-Pirjol, B. S., Cadar, E., Sirbu, R. & Negreanu-Pirjol, T. (2021). Antioxidant Activity of Some Fluids Extracts of Indigenous Wild

Cherry Fruits, *European Journal of Medicine and Natural Sciences*, *5*(1), 8-16

Negreanu-Pirjol, T., Negreanu-Pirjol, B. S., Popescu, A., Bratu, M. M., Udrea, M. & Busuricu, F. (2014). Comparative Antioxidant Properties of some Romanian Foods Fruits Extracts, *Journal of Environmental Protection and Ecology*, vol. *15*, no. 3, pp. 1139 – 1148.

Negreanu-Pirjol, T., Sirbu, R. & Negreanu-Pirjol, B.S. (2015). Antioxidant Activity of some Nutraceuticals based on Romanian Black and Red Fruits Mixed Extracts, *5th International Conference on Social Sciences - ICSS-2015*, Vol. *III*, pp. 253 – 262.

Olmedilla, B., Granado, F., Blanco, I. & Gil-Martínez, E. (1998). Carotenoid content in fruit and vegetables and its relevance to human health: Some of the factors involved. *Recent Research in Agricultural and Food Chemistry*, *2*, 57-70.

O'Sullivan, A. M., O'Callaghan, Y. C., O'Connor, T. P. & O'Brien, N. M. (2011). *Proceedings of Nutrition Society*, *70* (E61), doi: 10.1017/S0029665111001017.

Pal, R. S., Kumar, R. A., Agrawal P. K. & Bhatt, J. C. (2013). Antioxidant capacity and related phytochemicals analysis of methanolic extract of two wild edible fruits from North Western Indian Himalaya. *International Journal of Pharma and Bio Sciences*, *4* (2), 113-123.

Pérez-Lainez, M. D., Corona-Torres, T., García-Mateos, M. R., Winkler, R., Barrientos-Priego, A. F., Nieto-Ángel, R., Aguilar-Rincón, V. H. & García-Velázquez, J. A. (2019). Metabolomic study of volatile compounds in the pigmented fruit from Mexico *Crataegus* genotypes. *Journal of Applied Botany and Food Quality*, *92*, 15-23.

Perkins-Veazie, P., Fernandez, G. E. & Ma, G. (2020). Postharvest shelf life and composition of black raspberry selections and cultivars. *Acta Horticulturae*, *1277*, 455-460.

Perry, A., Rasmussen, H. & Johnson, E. J. (2009). Xanthophyll (lutein, zeaxanthin) content in fruits, vegetables and corn and egg products. *Journal of Food Composition and Analysis*, *22*, 9-15.

Petkova, N. T., Ognyanov, M. H., Vrancheva, R. Z. & Zhelev, P. (2020). Phytochemical, nutritional and antioxidant characteristics of whitebeam (*Sorbus aria*) fruits. *Acta Scientiarum Polonorum. Technologia Alimentaria, 19* (2), 219-229.

Ponder, A. & Hallman, E. (2017). Comparative evaluation of the nutritional value and the content of bioactive compounds in the fruit of individual species of chaenomeles and quince. *World Scientific News, 73*, 100-107.

Popoviciu, D. R. & Negreanu-Pirjol, T. (2019). Carotenoid, flavonoid and total phenolic content of *Sorbus torminalis* fruits. *Romanian-Arabian Journal of Geo-Bio-Diversity, 8* (1), 20-25.

Popoviciu, D. R., Negreanu-Pirjol, T. & Bercu, R. (2019). Carotenoids, flavonoids and total phenolic compounds concentration in fruits of milkflower cotoneaster (*Cotoneaster lacteus* W.W.Sm.). *Annals of the University of Craiova. Series Biology, Horticulture, Food Products Processing Technology, Environmental Engineering, 24* (60), 476-481.

Popoviciu, D. R., Negreanu-Pirjol, T. & Bercu, R. (2020a). Total carotenoid, flavonoid and phenolic compounds concentration in willowleaf cotoneaster (*Cotoneaster salicifolius* Franch.) fruits. *European Journal of Medicine and Natural Sciences, 4* (3), 1-6.

Popoviciu, D. R., Negreanu-Pirjol, T., Motelica, L. & Negreanu-Pirjol, B. (2020b). Carotenoids, flavonoids, total phenolic compounds and antioxidant activity of two creeping *Cotoneaster* species fruits extracts. *Revista de Chimie (Bucharest), 71* (3), 136-142.

Popoviciu, D. R., Negreanu-Pirjol, T., Motelica, L. & Negreanu-Pirjol, B. (2020c). Carotenoids, flavonoids, total phenolic compounds content and antioxidant activity of indigenous *Pyracantha coccinea* M. Roem. fruits. *Revista de Chimie (Bucharest), 71* (4), 258-266.

Pritwani, R. & Mathur, P. (2017). β-carotene content of some commonly consumed vegetables and fruits available in Delhi, India. *Journal of Nutrition & Food Sciences, 7* (5), doi: 10.4172/2155-9600.1000625.

Radenkovs, V. & Feldmane, D. (2017). Profile of lipophilic antioxidants in the by-products recovered from six cultivars of sour cherry (*Prunus cerasus* L.). *Natural Product Research, 31* (21), 2549-2553.

Serradilla, M.J., Hernández, A., López-Corrales, M., Ruiz-Moyano, S., Guía Córdoba, M. & Martín, A., (2016). *Nutritional Composition of*

Fruit Cultivars. Chapter 6 - *Composition of the Cherry* (*Prunus avium* L. and *Prunus cerasus* L.; Rosaceae), 127-147

Shameh, S., Alirezalu A., Hosseini, B. & Maleki, R. (2019). Fruit phytochemical composition and color parameters of 21 accessions of five *Rosa* species grown in North West Iran. *Journal of the Science of Food and Agriculture*, *99*, 5740-5751.

Solovchenko, A. E., Chivkunova, O. B., Merzlyak, M. N. & Gudkovsky, V. A. (2005). Relationships between chlorophyll and carotenoid pigments during on- and off-tree ripening of apple fruit as revealed non-destructively with reflectance spectroscopy. *Postharvest Biology and Technology*, *38*, 9-17.

Srednicka-Tober, D., Ponder, A., Hallmann, E., Głowacka, A. & Rozpara, E., (2019). The Profile and Content of Polyphenols and Carotenoids in Local and Commercial Sweet Cherry Fruits (*Prunus avium* L.) and Their Antioxidant Activity *In Vitro*. *Antioxidants*, *8*, 534.

Teixeira, M., Altmayer, T., Bruxel, F., Orlandi, C. R., de Moura, N. F., Afonso, C. N., Ethur, E. M., Hoehne, L. & de Freitas, E. M. (2018). *Rubus sellowii* Cham. & Schlitdl. (Rosaceae) fruit nutritional potential characterization. *Brazilian Journal of Biology*, *79* (3), 510-515.

Vinha, A. F., Machado, M., Santos, A. & Oliveira, M. B. P. P. (2012). Study of the influences by geographical origin in chemical characters, sugars, and antioxidant activity of Portuguese autochthonous *Prunus armeniaca* L. *Experimental Agriculture & Horticulture*, *1*, 8-20.

Vinholes, J., Gelain, D. P. & Vizzotto, M. (2016). Stone fruits as a source of bioactive compounds. In: da Silva, L.R., Silva, B.M. (eds.), *Natural Bioactive Compounds from Fruits and Vegetables as Health Promoters. Part I*, Bentham Science Publishers, Sharjah, UAE: 110-142.

Yoon, G. A., Yeum, K. J., Cho, Y. S., Chen, C. Y. O., Tang, G., Blumberg, J. B., Russell, R. M., Yoon, S. & Lee-Kim, Y. C. (2012). Carotenoids and total phenolic contents in plant foods commonly consumed in Korea. *Nutrition Research and Practice*, *6* (6), 481-490.

Zakynthinos, G. & Varzakas, T. (2016). Carotenoids: From Plants to Food Industry. *Current Research in Nutrition and Food Science*, *4* (Special Issue 1), 38-51.

Zechmeister, L. & Schroeder, W. A. (1942). The fruit of *Pyracantha angustifolia*: a practical source of pro-γ-carotene and prolycopene. *Journal of Biological Chemistry*, *144*, 315-320.

Zia-ul-Haq, M., Riaz, M., De Feo, V., Jaafar, H. Z. E. & Moga, M. (2014). *Rubus fruticosus* L.: constituents, biological activities and health related uses. *Molecules*, *19*, 10998-11029.

Zymonė, K., Raudonė, L., Raudonis, R., Marksa, M., Ivanauskas, L. & Janulis, V. (2018). Phytochemical profiling of fruit powders of twenty *Sorbus* L. cultivars. *Molecules*, *23*, 2593-2609.

BIOGRAPHICAL SKETCHES

B. S. Negreanu-Pirjol

Affiliation: Ovidius University of Constanta, Romania, Faculty of Pharmacy

Education: Pharmacist, Chemist, PhD

Research and Professional Experience: Head of disciplines Analytical Chemistry and Instrumental Analyse, respectively Instrumental Analyses Techniques – specialization Pharmacy and Pharmacy Assistance, Department of Pharmaceutical Sciences no. I, Faculty of Pharmacy, Ovidius University of Constanta, Romania

Professional Appointments: Associate Professor

Hirsch Index: 9 (ISI Web of Science / Clarivate Analytics / Publons); 6 (Scopus / Mendley); 9 (Scholar Google)

Honors:

Member in International Editorial (Advisory) Board:

1. Member in Editorial Board at *European Journal of Natural Sciences and Medicine*, https://journals.euser.org/index.php/ejmn/about/editorialTeam

Scientific Awards and Medals Obtained for Papers Presented in Scientific Events:

1. Gold Medal - la the 25th International Exhibition of Inventions "INVENTICA 2021", online, 23-25 June 2021, Iasi, Romania, for Romanian Complex Project no. 85PCCDI/2018, code PN-III-P1-1.2- PCCDI-2017-0701, title: *Complex Valorisation Of Black Sea Region Bioresources By Developing And Applying Innovative And Emerging Biotechnologies*
2. Silver Medal – at the 25th International Exhibition of Inventions "INVENTICA 2021", online, 23-25 June 2021, Iasi, Romania, for Patent no. 127726/29.11.2013, title: *Pharmaceutical Preparations Type Bioadhesive Gels Based On Chlorhexidine Metal Complexes And Process For Obtaining Them*
3. Bronse Medal - at the 25th International Exhibition of Inventions "INVENTICA 2021", online, 23-25 June 2021, Iasi, Romania, for Patent no. 126038/ 30.03.2012, title: *Ecological Fertilizer Biocomposit And Process For Obtaining It*
4. 3rd Prize, at International Symposium "Alternative and complementary therapies (Homeopathy / Phytotherapy)", online, 26 - 27 March 2021, Constanța, Romania
5. Mention, at International Symposium "Alternative and complementary therapies (Homeopathy / Phytotherapy)", online, 26 – 27 March 2021, Constanța, Romania
6. 1st Prize, at Workshop *"Pharmacy Past Present And Future"*- Fourth Edition, Ovidius University of Constanta, 16th-18th May

2019, Constanta, Romania, Program, Posters Session, paper title: *Synthesis, Characterization and Biological Activity of Some 3d And 4f Metal Complexes With Aromatic Biguanide Derivatives*,
7. 3rd Prize, at Workshop *"Pharmacy Past Present And Future"*- Fourth Edition, Ovidius University of Constanta, 16th-18th May 2019, Constanta, Romania, Program, Posters Session, paper title: *Synthesis, Characterization And Biological Activity Of Some 3d And 4f Metal Complexes With Aromatic Biguanide Derivatives*
8. Mention, at Workshop *"Pharmacy Past Present and Future"*- Third Edition, Ovidius University of Constanţa, 24th-25th May 2018, Constanta, Romania, Program, Posters Session, paper title: *Characterisation of marine algae, from Black Sea with modern analitical methods*
9. Awards, at International Conference *GLOREP 2018*, 15-17 November 2018, Timisoara, Romania, paper title: *Marine Biomass Valorisation as Potential Bioresourse for Biocosmetics and Eco-Agriculture*

Publications from the Last 3 Years:

RO-Patents

1. Negreanu-Pirjol, T; Negreanu-Pirjol, BS; Roncea, FN; Berger, DC; Moldovan, L; Rosioru, DM; Gaspar-Pintiliescu, A; Mitran, RA; Ranca, AM; Stefan, LM; Stanciuc (Prelipcean), AM; Matei, C; Bratu, MM; Lepadatu, AC; Artem, V; Erimia, CL; Vasile, M; *Topical dermato-cosmetic preparations type bioadhesive gels with dermal tissue regeneration effect*, Patent request RO 134938 A0, published in BOPI/28.05.2021
2. Negreanu-Pirjol, T; Negreanu-Pirjol, BS; Ranca, AM; Artem, V; Berger, DC; Moldovan, L; Rosioru, DM; Gaspar-Pintiliescu, A; Mitran, RA; Coroiu, V; Matei, C; Roncea, FN; Bratu, MM; Lepadatu, AC; Paraschiv, GM; Moise, I; Erimia, CL; Vasile, M; *Biostimulating-regenerating composition based on residual bioresources with*

fertilization potential, Patent request RO 134968 A0, published in BOPI/28.05.2021

3. Nastac, M; Negreanu-Pirjol, BS; Negreanu-Pirjol, T; Meghea, A; Gheorghiu, KA; Resteanu, AN; *Multicomposite biological fertilizer*, Patent RO131272 A2, published in BOPI/07.2016, p. 22

Articles Published in Publications Quoted ISI and BDI

[1] Bratu, MM; Birghila, S; Popescu, A; Negreanu-Pirjol, BS; Radu, M; Birghila, C. "Influence of Packaging Material on Polyphenol Content and Antioxidant Activity in Some Commercial Beers", *Processes*, 2021, 9(4), pp. 620-631.

[2] Artem, V; Negreanu–Pirjol, T; Ranca, A; Ciobanu, C; Abduraman, A; Coroiu, V; Negreanu-Pirjol, BS. „Experimental Studies On The Residual Marine And Viticultural Bioresources Valorization For New Organic Fertilizers", *U.P.B. Sci. Bull., Series B*, 2021, Vol. 83, Iss. 2, pp. 65-76.

[3] Negreanu-Pîrjol, T; Sirbu, R; Negreanu-Pirjol, BS; Cadar, E; Popoviciu, DR. "Preliminary Data Regarding Total Chlorophylls, Carotenoids and Flavonoids Content in *Flavoparmelia caperata* (L.) Hale Lichens Species", *European Journal of Medicine and Natural Sciences*, (2021), Vol. 5, Issue 1, pp. 1-7.

[4] Negreanu-Pîrjol, BS; Cadar, E; Sirbu, R; Negreanu-Pirjol, T. "Antioxidant Activity of Some Fluids Extracts of Indigenous Wild Cherry Fruits", *European Journal of Medicine and Natural Sciences*, (2021), Vol. 5, Issue 1, pp. 8-16.

[5] Sirbu, R; Negreanu-Pirjol, T; Negreanu-Pirjol, BS; Cadar, E. "Important Properties of Grapes and Wine from the Dobrogea Area for Therapeutic Use", *European Journal of Medicine and Natural Sciences*, (2021), Vol. 5, Issue 1, pp. 17-26.

[6] Cadar, E; Negreanu-Pirjol, BS; Negreanu-Pirjol, T; Sirbu, R. "Characteristics of Techirghiol Sludge and Different Methods of Peloid Therapy", *European Journal of Medicine and Natural Sciences,* (2021), Vol. 5, Issue 1, pp. 27-35.

[7] Popoviciu, DR; Negreanu-Pirjol, T; Motelica, L; Negreanu-Pirjol, BS. "Carotenoids, Flavonoids, Total Phenolic Compounds Content and Antioxidant Activity of Indigenous *Pyracantha coccinea* M. Roem. Fruits", *Revista de Chimie (Bucharest)*, 71(4), 2020, 258-266.

[8] Negreanu-Pîrjol, T; Negreanu-Pîrjol, BS; Popoviciu, DR; Roncea, FN. "Preliminary Data Regarding Pharmaceutical Forms Type Gels Based on Marine Algae Extracts with Antioxidant Activity", *European Journal of Medicine and Natural Sciences*, (2020), Vol. 4, Issue 3, pp. 55-65.

[9] Popoviciu, DR; Negreanu-Pirjol, T; Motelica, L; Negreanu-Pirjol, BS. "Carotenoids, Flavonoids, Total Phenolic Compounds and Antioxidant Activity of Two Creeping *Cotoneaster* Species Fruits Extracts", *Revista de Chimie (Bucharest)*, 71(3), 2020, 136-142.

[10] Negreanu-Pirjol, BS; Negreanu-Pirjol, T; Paraschiv, GM; Bratu, MM. "Residual Marine Algae Biomass - An Important Raw Material for Obtaining a Soil Biostimulator-Regenerator", *European Journal of Medicine and Natural Sciences*, (2020), Vol. 4, Issue 3, pp.74-85.

[11] Sirbu, R; Negreanu-Pirjol, T; Mirea, M; Negreanu-Pirjol, BS. "Bioactive Compounds from Three Green Algae Species along Romanian Black Sea Coast with Therapeutically Properties", *European Journal of Medicine and Natural Sciences*, (2019), Vol. 3, No. 1, pp. 5 -15.

[12] Negreanu-Pirjol, T; Sirbu, R; Mirea, M; Negreanu-Pirjol, BS. "Antioxidant activity correlated with chlorophyll pigments and magnesium content of some green seaweeds", *European Journal of Medicine and Natural Sciences,* (2019), Vol. 3, No. 1, pp. 16 – 22.

[13] Negreanu-Pirjol, BS; Negreanu-Pirjol, T; Sirbu, R; Popoviciu, DR. "Bioaccumulation and Effects of Aluminium on Plant Growth in Three Culture Plants Species", *Revista de Chimie (Bucharest)*, 2019, vol. 70, nr. 2, pp. 602 – 604. [*Journal of Chemistry*]

[14] Cadar, E; Sirbu, Pirjol R; Negreanu, BS; Ionescu, AM; Pirjol, T. Negreanu, "Heavy Metals Bioaccumulation Capacity on Marine Algae Biomass from Romanian Black Sea Coast", *Revista de*

Chimie (Bucharest), vol. 70, nr. 8, 2019, pp. 3065 – 3072. [*Journal of Chemistry*]

[15] Negreanu-Pîrjol, T; Negreanu-Pîrjol, BS; Popoviciu, DR; Mirea M., Vasile, M. "Micro-, macroelements and heavy metals in the algae component of a new biostimulator-regenerator for grapevine soils", 19th International Multidisciplinary Scientific GeoConferences– *SGEM*, 2019, 9–11 December 2019, Vienna, Austria, *Conference Proceedings, Volume 19 - Nano, Bio, Green and Space: Technologies for a Sustainable Future*, Issue 6.3, Section "Advances in Biotechnology", paper 19, pp. 141-148.

[16] Bratu, MM; Birghila, S; Popescu, A; Negreanu-Pirjol, BS; Negreanu-Pirjol, T. "Correlation of antioxidant activity of dried berry infusions with the polyphenols and selected microelements contents", *Bulletin of the Chemical Society of Ethiopia*, 2018, Vol. 32, No. 1, pp. 1-12.

[17] Negreanu-Pirjol, BS; Negreanu-Pirjol, T; Mirea, M; Vasile, M; Cadar, E. "Antioxidant capacity of some marine green macroalgae species fluid extracts", 18th International Multidisciplinary Scientific Conferences on Earth & GeoSciences – *SGEM Vienna Green 2018*, 3 – 6 December 2018, Vienna, Austria, Hofburg Congress Center, *Conference Proceedings, Volume 18 - Nano, Bio, Green and Space Technologies for a Sustainable Future*, Issue 6.4, Section 8 "Advances in Biotechnology", paper 51, pp. 63 – 69.

[18] Negreanu-Pirjol, BS; Negreanu-Pirjol, T; Rosioru, D; Chisoi, A; Aschie, M. "Microbiological characterization of the waste algal biomass along Romanian Black Sea Coast", *18th International Multidisciplinary Scientific Conferences on Earth & GeoSciences – SGEM Vienna Green 2018*, 3 – 6 December 2018, Vienna, Austria, Hofburg Congress Center, Conference Proceedings, Volume 18 - Nano, Bio, Green and Space Technologies for a Sustainable Future, Issue 6.4, Section 8 "Advances in Biotechnology", paper 51, pp. 211 – 217.

[19] Popoviciu, DR; Negreanu-Pîrjol, BS; Negreanu-Pîrjol, T. "Chromium bioaccumulation in three Poaceae species", *Annals of the University of Craiova*, Editura Universitaria, Craiova, 2018,

Series: *Biology, Horticulture, Food produce processing technology, Environmental engineering*, vol. XXIII (LIX), oct. 2018, pp. 462-466.

D. R. Popoviciu

Affiliation: Ovidius University of Constanta, Romania, Faculty of Natural Sciences and Agricultural Sciences

Education: Biologist, PhD

Research and Professional Experience: Assistant professor, Faculty of Natural Sciences and Agricultural Sciences, Ovidius University of Constanta, Romania

Professional Appointments: Assistant Professor

Publications from the Last 3 Years:

Articles Published in Publications Quoted ISI and BDI

1. Negreanu-Pirjol, T; Sirbu, R; Negreanu-Pirjol, BS; Cadar, E; Popoviciu, DR. "Preliminary Data Regarding Total Chlorophylls, Carotenoids and Flavonoids Content in *Flavoparmelia caperata* (L.) Hale Lichens Species", *European Journal of Medicine and Natural Sciences*, (2021), Vol. 5, Issue 1, pp. 1-7.
2. Negreanu-Pirjol, T; Negreanu-Pirjol, BS; Popoviciu, DR; Roncea, FN. "Preliminary Data Regarding Pharmaceutical Forms Type Gels Based on Marine Algae Extracts with Antioxidant Activity", *European Journal of Medicine and Natural Sciences*, (2020), Vol. 4, Issue 3, pp. 55-65.
3. Popoviciu, DR; Negreanu-Pirjol, T; Bercu, R. "Total Carotenoid, Flavonoid and Phenolic Compounds Concentration in Willowleaf

Cotoneaster (*Cotoneaster salicifolius* Franch.) Fruits", *European Journal of Medicine and Natural Sciences*, (2020), Vol. 4, Issue 3, pp. 1 – 6.
4. Popoviciu, DR; Negreanu-Pirjol, T; Motelica, L; Negreanu-Pirjol BS. "Carotenoids, Flavonoids, Total Phenolic Compounds Content and Antioxidant Activity of Indigenous *Pyracantha coccinea* M. Roem. Fruits", *Revista de Chimie (Bucharest)*, 71(4), 2020, 258-266.
5. Popoviciu, DR; Negreanu-Pirjol, T; Motelica, L; Negreanu-Pirjol, BS. "Carotenoids, Flavonoids, Total Phenolic Compounds and Antioxidant Activity of Two Creeping *Cotoneaster* Species Fruits Extracts", *Revista de Chimie (Bucharest)*, 71(3), 2020, 136-142. [*Journal of Chemistry*]
6. Negreanu-Pirjol, BS; Negreanu-Pirjol, T; Sirbu, R; Popoviciu, DR. "Bioaccumulation and Effects of Aluminium on Plant Growth in Three Culture Plants Species", *Revista de Chimie (Bucharest)*, vol. 70, nr. 2, 2019, pp. 602 – 604. [*Journal of Chemistry*]
7. Negreanu-Pirjol, T; Negreanu-Pirjol, BS; Popoviciu, DR; Mirea, M; Vasile, M. "Micro-, macroelements and heavy metals in the algae component of a new biostimulator-regenerator for grapevine soils", 19th International Multidisciplinary Scientific GeoConferences–*SGEM 2019*, 9–11 December 2019, Vienna, Austria, *Conference Proceedings, Volume 19 - Nano, Bio, Green and Space: Technologies for a Sustainable Future*, Issue 6.3, Section "Advances in Biotechnology", paper 19, pp. 141-148.
8. Popoviciu, DR; Negreanu-Pirjol, T; Bercu, R. "Carotenoids, flavonoids and total phenolic compounds concentration in fruits of Milkflower Cotoneaster (*Cotoneaster lacteus* W. W. Sm.)", *Annals of the University of Craiova*, Editura Universitaria, Craiova, 2019, *Series: Horticulture, Food products processing technology*, vol. XXIV (LX), oct. 2019, pp. 476 – 481.
9. Popoviciu, DR; Negreanu-Pirjol, BS; Negreanu-Pirjol, T. "Chromium bioaccumulation in three *Poaceae* species", *Annals of the University of Craiova*, Editura Universitaria, Craiova, 2018, *Series: Biology*,

Horticulture, Food produce processing technology, Environmental engineering, vol. XXIII (LIX), Oct. 2018, pp. 462-466.
10. Popoviciu, DR; Negreanu-Pirjol, T. "Copper, manganese and zinc bioaccumulation in three common woody species from Black Sea coastal area", *U.P.B. Sci. Bull., Series B*, Vol. 80, Iss. 2, 2018, pp. 49-56.
11. Popoviciu, DR; Negreanu-Pirjol, T; Sirbu, "Copper tolerance and bioaccumulation in seedlings of white mustard (*Sinapis alba* L.)", *Journal of Science and Arts, Chemistry Section*, Year 18, No. 1(42), pp. 211-218, 2018.
12. Popoviciu, DR; Bercu, R; Negreanu-Pirjol, T. "Chromium bioaccumulation in three common woody species", *Annals of the University of Craiova*, Editura Universitaria, Craiova, 2018, *Series: Biology, Horticulture, Food produce processing technology, Environmental engineering*, vol. XXIII (LIX), oct. 2018, p. 467-471.

T. Negreanu-Pirjol

Affiliation: Ovidius University of Constanta, Romania, Faculty of Pharmacy

Education: Pharmacist, Chemist, PhD

Research and Professional Experience: Head of disciplines General and Inorganic Chemistry - specialization Pharmacy and Pharmacy Assistance, respectively Parapharmaceuticals and Technical-Medical Products – specialization Pharmacy Assistance, Department of Pharmaceutical Sciences no. I, Faculty of Pharmacy, Ovidius University of Constanta, Romania

Professional Appointments: Professor

Hirsch Index: 11 (ISI Web of Science / Clarivate Analytics / Publons); 11 (Scopus / Mendley); 12 (Scholar Google).

Honors:

- Departmental Coordinator for the Erasmus Plus Mobility Program at the Faculty of Pharmacy, "Ovidius" University of Constanta, Romania, confirmation of appointment by Decision of the Council of the Faculty of Pharmacy.
- Member of the Research-Development-Innovation Council of the "Ovidius" University of Constanta, Romania.

Reviewer and Member in International Editorial (Advisory) Board

1. Reviewer, since 2021, at *Issues in Biological Sciences and Pharmaceutical Research*, ISSN 2350-1588, https://www.journalissues.org/ IBSPR/
2. Reviewer, since 2015, at *U.P.B. Sci. Bull, Series B,* ISSN 1454-2331, https://www.scientificbulletin.upb.ro/SeriaB_-_Chimie_si_Stiinta_Materialelor.php?page=recenzare
3. Reviewer, for 2014, at *Journal of Environmental Protection and Ecology*, ISSN 1311-5065.
4. Member in Editorial Team, *Revista Cercetări Marine - Revue Recherches Marines - Marine Research Journal*, http://www.marine-research-journal.org/index.php/cmrm/about/editorialTeam, since 2019
5. Member in International Editorial and Advisory Board la Revista *European Journal of Natural Sciences and Medicine*, since 2019, Vol. 1, No.1 (2018): EJNM, January-April 2018, DOI: http://dx.doi.org/10.26417/ejnm.v1i1, ISSN 2601-8691 (Print), ISSN 2601-8705 (Online); Vol 1, No. 2 (2018): EJNM, May-August 2018, DOI: http://dx.doi.org/10.26417/ejnm.v1i2, ISSN 2601-8691 (Print), ISSN 2601-8705 (Online) http://journals.euser.org/index.php/ejnm/

6. Member in International Editorial and Advisory Board la Revista *European Journal of Medicine and Natural Sciences*, Vol. 1, No. 1, (2017), EJMN, September-December 2017, DOI: http://dx.doi.org/10.26417/ejmn.v1i1, ISSN 2601-6397 (Print), ISSN 2601-6400 (Online); Vol. 2, No.1 (2018), EJMN, January-April 2018, DOI: http://dx.doi.org/10.26417/ejmn.v2i1, ISSN 2601-6397 (Print), ISSN 2601-6400 (Online) http://journals.euser.org/index.php/ejmn/.
7. Member in Advisory Board (Editorial Board) la Revista *Scripta Scientifica Pharmaceutica*, Medical University of Varna, Bulgaria, since 2015, http://mu-varna.bg/EN/Publishing/Documents/Editorial%20Board.pdf; http://journals.mu-varna.bg/index.php/ssp/about/editorialTeam.

Scientific Awards and Medals Obtained for Papers Presented in Scientific Events:

1. Gold Medal - at the 25th International Exhibition of Inventions "INVENTICA 2021", online, 23-25 June 2021, Iasi, Romania, for Romanian Complex Project no. 85PCCDI/2018, code PN-III-P1-1.2-PCCDI-2017-0701, title: *Complex Valorisation Of Black Sea Region Bioresources By Developing And Applying Innovative And Emerging Biotechnologies*
2. Silver Medal - at the 25th International Exhibition of Inventions "INVENTICA 2021", online, 23-25 June 2021, Iasi, Romania, for Patent no. 127726/29.11.2013, title: *Pharmaceutical Preparations Type Bioadhesive Gels Based On Chlorhexidine Metal Complexes And Process For Obtaining Them*

3. Bronse Medal - at the 25th International Exhibition of Inventions "INVENTICA 2021", online, 23-25 June 2021, Iasi, Romania, for Patent no. 126038/ 30.03.2012, title: *Ecological Fertilizer Biocomposit And Process For Obtaining It*
4. 3rd Prize, at International Symposium *Alternative and complementary therapies (Homeopathy / Phytotherapy)*, online, 26 - 27 March 2021, Constanta, Romania
5. 2nd Prize, at International Symposium *Alternative and complementary therapies (Homeopathy / Phytotherapy)*, online, 26 - 27 March 2021, Constanta, Romania
6. Mention, at International Symposium *Alternative and complementary therapies (Homeopathy / Phytotherapy)*, online, 26 - 27 March 2021, Constanta, Romania
7. 1st Prize, at Workshop *"Pharmacy Past Present And Future"*- Fourth Edition, Ovidius University of Constanța, 16th-18th May 2019, Constanta, Romania, Program, Posters Session, paper title: *Synthesis, Characterization and Biological Activity of Some 3d And 4f Metal Complexes With Aromatic Biguanide Derivatives*
8. 3rd Prize, at Workshop *"Pharmacy Past Present And Future"*- Fourth Edition, Ovidius University of Constanta, 16th-18th May 2019, Constanta, Romania, Program, Posters Session, paper title: *Synthesis, Characterization And Biological Activity Of Some 3d And 4f Metal Complexes With Aromatic Biguanide Derivatives*
9. Mention, at Workshop *"Pharmacy Past Present And Future"*- Third Edition, Ovidius University of Constanța, 24th-25th May 2018, Constanta, Romania, Program, Posters Session, paper title: *Characterisation of marine algae, from Black Sea with modern analitical methods*
10. Awards, at Intenational Conference *GLOREP 2018*, 15-17 November 2018, Timisoara, Romania, paper title: *Marine Biomass Valorisation as Potential Bioresourse for Biocosmetics and Eco-Agriculture*

Publications from the Last 3 Years:

RO-Patents

1. Negreanu-Pirjol, T; Negreanu-Pirjol, BS; Roncea, FN; Berger, DC; Moldovan, L; Rosioru, DM; Gaspar-Pintiliescu, A; Mitran, RA; Ranca, AM; Stefan, LM; Stanciuc (Prelipcean), AM; Matei, C; Bratu, MM; Lepadatu, AC; Artem, V; Erimia, CL; Vasile, M; *Topical dermato-cosmetic preparations type bioadhesive gels with dermal tissue regeneration effect*, Patent request RO 134938 A0, published in BOPI/28.05.2021
2. Negreanu-Pirjol, T; Negreanu-Pirjol, BS; Ranca, AM; Artem, V; Berger, DC; Moldovan, L; Rosioru, DM; Gaspar-Pintiliescu, A; Mitran, RA; Coroiu, V; Matei, C; Roncea, FN; Bratu, MM; Lepadatu, AC; Paraschiv, GM; Moise, I; Erimia, CL; Vasile, M; *Biostimulating-regenerating composition based on residual bioresources with fertilization potential*, Patent request RO 134968 A0, published in BOPI/28.05.2021
3. Moldovan, L; Gaspar-Pintiliescu, A; Crăciunescu, O; Ştefan, LM; Stanciuc, AM; Coroiu, V; Negreanu-Pirjol T. *Process for obtaining gelatin with bioactive properties from marine gastropods*, Patent request RO134119 (A0), published in BOPI/29.05.2020.
4. Nastac, M; Negreanu-Pirjol, BS; Negreanu-Pirjol, T; Meghea, A; Gheorghiu, KA; Resteanu, AN; *Multicomposite biological fertilizer*, Patent RO131272 A2, published in BOPI/07.2016, p. 22

Chapter and Articles Published in Publications Quoted ISI and BDI

[1] Bratu, MM; Negreanu–Pirjol, T. "Food and Beverage Consumption and Health. Fruit and Pomace Extracts – Biological activity, Potential Application and Beneficial Health Effects", Chapter 12, *Elderberries Extracts: Biologic Effects, Application for Therapy: A Review*, Jason P. Owen Editor, Nova Science Publishers, Inc., New York, Future Medicine Ltd., 2015, p. 227–240.

[2] Artem, V; Negreanu–Pirjol, T; Ranca, A; Ciobanu, C; Abduraman, A; Coroiu, V; Negreanu-Pirjol, BS. „Experimental Studies On The Residual Marine And Viticultural Bioresources Valorization For New Organic Fertilizers", *U.P.B. Sci. Bull., Series B*, 2021, Vol. 83, Iss. 2, pp. 65-76.

[3] Brezoiu, AM; Bajenaru, L; Berger, D; Mitran, RA; Deaconu, M; Lincu, D; Stoica Guzun, A; Matei, C; Georgeta Moisescu, M; Negreanu-Pirjol, T. "Effect of Nanoconfinement of Polyphenolic Extract from Grape Pomace into Functionalized Mesoporous Silica on Its Biocompatibility and Radical Scavenging Activity", *Antioxidants*, (MDPI), 2020, 9(8), 696.

[4] Brezoiu, AM; Prundeanu, M; Berger, D; Deaconu, M; Matei, C; Oprea, O; Vasile, E; Negreanu-Pîrjol, T; Muntean, D; Danciu, C. "Properties of *Salvia offcinalis* L. and *Thymus serpyllum* L. extracts free and embedded into mesopores of silica and titania", *Nanomaterials*, (MDPI), 2020, Vol. 10, Iss. 5, 820.

[5] Negreanu-Pirjol, BS; Negreanu-Pirjol, T; Sirbu, R; Popoviciu, DR. "Bioaccumulation and Effects of Aluminium on Plant Growth in Three Culture Plants Species", *Revista de Chimie (Bucharest)*, vol. 70, nr. 2, 2019, pp. 602–604. [*Journal of Chemistry*]

[6] Cadar, E; Sirbu, R; Pirjol, BS; Negreanu, AM; Ionescu, Pirjol T. Negreanu. "Heavy Metals Bioaccumulation Capacity on Marine Algae Biomass from Romanian Black Sea Coast", *Revista de Chimie (Bucharest)*, vol. 70, nr. 8, 2019, pp. 3065–3072. [*Journal of Chemistry*]

[7] Biris-Dorhoi, ES; Tofana, M; Chis, SM; Lupu, CE; Negreanu-Pirjol, T. "Wastewater treatment using marine algae biomass as pollutants removal", *Revista de Chimie (Bucharest)*, vol. 69, No. 5, 2018, pp. 1089-1098. [*Journal of Chemistry*]

[8] Biris-Dorhoi, SE; Tofana, M; Popoviciu, DR; Negreanu-Pirjol, T. "Oxidative stress evaluation in organic pollution conditions on some marine algae species", *Journal of Environmental Protection and Ecology*, vol. 19, no. 2 (2018), pp. 592-600.

[9] Bratu, MM; Birghila, S; Popescu, A; Negreanu-Pirjol, BS; Negreanu-Pirjol, T. "Correlation of antioxidant activity of dried berry infusions with the polyphenols and selected microelements contents", *Bulletin of the Chemical Society of Ethiopia*, 2018, Vol. 32, No. 1, pp. 1-12.

[10] Calinescu, M; Fiastru, M; Bala, D; Mihailciuc, C; Negreanu-Pirjol, T; Jurca, B. "Synthesis, characterization, electrochemical behavior and antioxidant activity of new copper(II) coordination compounds with curcumin derivatives", *Journal of Saudi Chemical Society*, King Saud University, Production and hosting by Elsevier B.V, (noiembrie 2019), vol. 23, Iss. 7, p. 817-827.

[11] Gaspar-Pintiliescu, A; Stefan, LM; Anton, ED; Berger, D; Matei, C; Negreanu-Pirjol, T; Moldovan, L. "Physicochemical and Biological Properties of Gelatin Extracted from Marine Snail *Rapana venosa*", *Marine Drugs*, (MDPI), 2019, vol. 17, Iss. 10, pp. 589.

[12] Mihalache, M; Negreanu-Pirjol, T; Dumitrașcu, F; Drăghici, C; Călinescu, M. "Synthesis, characterization and biological activity of new Ni(II), Pd(II) and Cr(III) complex compounds with chlorhexidine", *Journal of the Serbian Chemical Society (J. Serb. Chem. Soc.)*, vol. 83, no. 3, pp. 271–284 (2018).

[13] Roșioru, DM; Oros, A; Coatu, V; Stoica, E; Negreanu-Pirjol, T. "Estimation of *Rapana venosa* (Valenciennes, 1846) quality, a marine living resource from the Romanian Black Sea with bioeconomic importance", *UPB - Scientific Bulletin (U.P.B. Sci. Bull.)*, Series B, Vol. 82, Iss. 2, 2020, pp. 39-46.

[14] Popoviciu, DR; Negreanu-Pirjol, T. "Copper, manganese and zinc bioaccumulation in three common woody species from Black Sea coastal area", *U.P.B. Sci. Bull., Series B*, Vol. 80, Iss. 2, 2018, pp. 49-56.

[15] Popoviciu, DR; Negreanu-Pirjol, T; Sirbu, R. "Copper tolerance and bioaccumulation in seedlings of white mustard (*Sinapis alba* L.)", *Journal of Science and Arts, Chemistry Section*, Year 18, No. 1(42), pp. 211-218, 2018.

[16] Negreanu-Pirjol, T; Negreanu-Pirjol, BS; Popoviciu, DR; Mirea, M; Vasile, M. "Micro-, macroelements and heavy metals in the algae

component of a new biostimulator-regenerator for grapevine soils", 19th International Multidisciplinary Scientific GeoConferences–*SGEM 2019*, 9–11 December 2019, Vienna, Austria, *Conference Proceedings, Volume 19 - Nano, Bio, Green and Space: Technologies for a Sustainable Future*, Issue 6.3, Section "Advances in Biotechnology", paper 19, pp. 141-148.

[17] Negreanu-Pirjol, BS; Negreanu-Pirjol, T; Mirea, M; Vasile, M; Cadar, E. "Antioxidant capacity of some marine green macroalgae species fluid extracts", *18th International Multidisciplinary Scientific Conferences on Earth & GeoSciences – SGEM Vienna Green 2018*, 3 – 6 December 2018, Vienna, Austria, Hofburg Congress Center, Conference Proceedings, Volume 18 - Nano, Bio, Green and Space Technologies for a Sustainable Future, Issue 6.4, Section 8 "Advances in Biotechnology", paper 51, pp. 63 – 69.

[18] Negreanu-Pîrjol, T; Sirbu, R; Negreanu-Pirjol, BS; Cadar, E; Popoviciu, DR. "Preliminary Data Regarding Total Chlorophylls, Carotenoids and Flavonoids Content in *Flavoparmelia caperata* (L.) Hale Lichens Species", *European Journal of Medicine and Natural Sciences*, 2021, Vol. 5, Iss. 1, pp. 1-7.

[19] Negreanu-Pirjol, T; Negreanu-Pirjol, BS; Popoviciu, DR; Roncea, FN. "Preliminary Data Regarding Pharmaceutical Forms Type Gels Based on Marine Algae Extracts with Antioxidant Activity", *European Journal of Medicine and Natural Sciences*, 2020, Vol. 4, Iss. 3, pp. 55-65.

[20] Negreanu-Pirjol, T; Sirbu, R; Mirea, M; Negreanu-Pirjol, BS. "Antioxidant activity correlated with chlorophyll pigments and magnesium content of some green seaweeds", *European Journal of Medicine and Natural Sciences,* 2019, Vol. 3, No. 1, pp. 16–22.

[21] Negreanu-Pîrjol, BS; Cadar, E; Sirbu, R; Negreanu-Pirjol, T. "Antioxidant Activity of Some Fluids Extracts of Indigenous Wild Cherry Fruits", *European Journal of Medicine and Natural Sciences*, 2021, Vol. 5, Iss. 1, pp. 8-16.

[22] Sirbu, R; Negreanu-Pirjol, T; Negreanu-Pirjol, BS; Cadar, E. "Important Properties of Grapes and Wine from the Dobrogea Area for Therapeutic Use", *European Journal of Medicine and Natural Sciences*, 2021, Vol. 5, Iss. 1, pp. 17-26.

[23] Popoviciu, DR; Negreanu-Pirjol, T; Motelica, L; Negreanu-Pirjol, BS. "Carotenoids, Flavonoids, Total Phenolic Compounds Content and Antioxidant Activity of Indigenous *Pyracantha coccinea* M. Roem. Fruits", *Revista de Chimie (Bucharest)*, 71(4), 2020, 258-266.

[24] Popoviciu, DR; Negreanu-Pirjol, T; Motelica, L; Negreanu-Pirjol, BS. "Carotenoids, Flavonoids, Total Phenolic Compounds and Antioxidant Activity of Two Creeping *Cotoneaster* Species Fruits Extracts", *Revista de Chimie (Bucharest)*, 71(3), 2020, 136-142.

[25] Negreanu-Pirjol, BS; Negreanu-Pirjol, T; Paraschiv, GM; Bratu, MM. "Residual Marine Algae Biomass - An Important Raw Material for Obtaining a Soil Biostimulator-Regenerator", *European Journal of Medicine and Natural Sciences,* 2020, Vol. 4, Iss. 3, pp. 74-85.

[26] Sirbu, R; Negreanu-Pirjol, T; Mirea, M; Negreanu-Pirjol, BS. "Bioactive Compounds from Three Green Algae Species along Romanian Black Sea Coast with Therapeutically Properties", *European Journal of Medicine and Natural Sciences,* 2019, Vol. 3, No. 1, pp. 5-15.

[27] Popoviciu, DR; Negreanu-Pirjol, BS; Negreanu-Pirjol, T. "Chromium bioaccumulation in three Poaceae species", *Annals of the University of Craiova*, Editura Universitaria, Craiova, 2018, *Series: Biology, Horticulture, Food produce processing technology, Environmental engineering*, vol. XXIII (LIX), 2018, pp. 462-466.

[28] Cadar, E; Negreanu-Pirjol, BS; Negreanu-Pirjol, T; Sirbu, R. "Characteristics of Techirghiol Sludge and Different Methods of Peloid Therapy", *European Journal of Medicine and Natural Sciences*, 2021, Vol. 5, Issue 1, pp. 27-35.

[29] Popoviciu, DR; Negreanu-Pirjol, T; Bercu, R. "Total Carotenoid, Flavonoid and Phenolic Compounds Concentration in Willowleaf Cotoneaster (*Cotoneaster salicifolius* Franch.) Fruits", *European Journal of Medicine and Natural Sciences,* 2020, Vol. 4, Iss. 3, pp. 1-6.

[30] Popoviciu, DR; Negreanu-Pirjol, T; Bercu, R. "Carotenoids, flavonoids and total phenolic compounds concentration in fruits of Milkflower Cotoneaster (*Cotoneaster lacteus* W.W.Sm.)", *Annals of the University of Craiova*, Editura Universitaria, Craiova, 2019, *Series: Horticulture, Food products processing technology*, vol. XXIV (LX), 2019, pp. 476–481.

INDEX

A

acetone, 5, 70, 92, 93
acid, 9, 34, 42, 48, 72, 73, 74, 103, 112, 115, 127
additives, ix, 14, 16, 68, 71
adults, 34, 53, 58, 59, 60, 63
age, 29, 34, 50, 54, 56, 61, 71
algae, vii, ix, 2, 33, 67, 70, 71, 103, 113, 142, 145, 147, 151, 153, 154
antioxidant, vii, viii, 2, 6, 7, 12, 13, 16, 17, 24, 34, 44, 50, 53, 57, 62, 65, 70, 71, 95, 98, 102, 114, 127, 128, 130, 131, 132, 133, 134, 135, 136, 137, 138, 139, 145, 154
apples, x, 111, 112, 119, 120, 121
aquaculture, 8, 9, 94
aqueous solutions, 93
arabinogalactan, 60
aromatic compounds, 115
astaxanthin, v, vii, 1, 2, 4, 5, 6, 7, 8, 9, 12, 13, 14, 15, 16, 17, 18, 19, 20, 21, 22, 23, 24, 73, 76, 78, 79, 85, 88, 89, 90, 91, 94, 95, 97, 100, 101, 102, 103, 107, 109, 115
atherosclerosis, 35, 54, 55

B

bacteria, vii, ix, 33, 67, 70, 90, 113
benefits, viii, 4, 7, 9, 10, 22, 27, 41, 68, 94, 95, 127
beta-carotene, 58, 108
bioaccessibility, viii, 28, 29, 35, 38, 39, 41, 42, 43, 44, 52, 55, 56, 57, 59, 62, 65, 115, 132
bioaccumulation, 145, 147, 148, 154, 156
bioactivity, viii, 2, 16, 17, 28, 130
bioavailability, vii, viii, 28, 29, 32, 35, 42, 43, 46, 47, 49, 50, 51, 52, 54, 58, 59, 60, 62, 63, 64, 65, 115, 130
biochemistry, 18
bioconversion, ix, 28, 29, 65
biodiesel, 75, 102
biodiversity, 104
biological activities, viii, 27, 130, 132, 140
biomass, 77, 83, 84, 87, 88, 92, 93, 96, 102, 108, 145, 153
biomolecules, ix, 68, 92
biosynthesis, ix, 18, 68, 72, 74, 75, 83, 86, 88, 90, 91, 97, 100, 103, 107, 109, 115, 116, 130, 131, 133

biosynthetic pathways, 73
biotechnology, 22, 74, 88, 131
bonds, viii, 2, 5, 27, 28, 70, 71, 115
by-products, 34, 103, 138

C

carbon, x, 5, 6, 10, 28, 68, 73, 74, 76, 77, 79, 80, 82, 95, 96, 113, 114
cardiovascular disease, 3, 29, 71, 95
carotene, viii, x, 2, 3, 4, 5, 6, 7, 13, 15, 28, 44, 46, 53, 55, 56, 58, 59, 63, 70, 73, 74, 75, 76, 77, 78, 79, 84, 85, 87, 89, 90, 91, 93, 94, 95, 97, 101, 102, 103, 105, 107, 109, 112, 113, 118, 119, 120, 121, 122, 124, 126, 127, 128, 129, 130, 138, 140
challenges, vii, ix, 10, 16, 28, 49, 62
chemical characteristics, 12, 13
chemical properties, 130, 133
chemical stability, 51, 53, 60, 65
chlorophyll, 3, 58, 61, 113, 114, 139, 144, 155
cholesterol, 34, 44, 60, 63
cluster of differentiation, 44
color, iv, viii, 27, 28, 71, 76, 78, 88, 94, 115, 127, 130, 135, 139
commercial, 7, 94, 115, 120, 123, 126, 133
composition, vii, 2, 50, 63, 69, 77, 80, 113, 122, 124, 132, 134, 135, 137, 139, 142, 152
compounds, vii, x, 2, 6, 11, 13, 14, 15, 16, 38, 41, 60, 69, 70, 76, 86, 88, 90, 93, 107, 112, 113, 119, 121, 122, 126, 127, 128, 130, 131, 132, 133, 135, 136, 137, 138, 139, 154
consumers, viii, ix, 2, 7, 68, 112, 131
consumption, viii, 6, 15, 16, 27, 29, 34, 35, 38, 42, 49, 57, 127, 135
cooking, ix, 12, 28, 36, 37, 39, 55, 60, 64, 66
cosmetic, 4, 11, 76, 94, 95, 115, 142, 152

cost, 9, 10, 33, 69, 72, 74, 76, 81, 91, 95, 100, 108
cultivars, 17, 112, 118, 120, 121, 122, 124, 125, 126, 127, 128, 132, 134, 135, 137, 138, 140
cultivation, ix, 68, 69, 71, 80, 83, 84, 86, 96, 99, 105, 109, 133
cultivation conditions, 71, 83, 96
culture, 69, 71, 74, 78, 80, 82, 83, 84, 85, 97, 107
culture medium, 69, 71, 74, 80, 82, 84

D

degradation, ix, 3, 10, 12, 15, 28, 38, 39, 41, 61
diet, 4, 6, 29, 30, 38, 42, 47, 55, 63, 70, 100, 127
dietary challenges, 28
dietary fat, 53
dietary fiber, 46, 59, 63
dietary intake, 29, 71
dietary supplementation, 54, 55, 58
digestion, viii, 28, 41, 42, 43, 45, 47, 49, 60
distribution, ix, 25, 50, 51, 68, 70, 75, 119
double bonds, viii, 2, 5, 7, 13, 15, 27, 29, 71, 72, 114
downstream steps, ix, 68, 91, 93

E

egg, 30, 32, 55, 57, 61, 63, 95, 137
engineering, 88, 89, 90, 91, 97, 100, 101, 102, 109, 146, 148, 156
extraction, 2, 9, 10, 11, 12, 17, 18, 19, 21, 22, 23, 24, 32, 33, 69, 71, 92, 93, 96, 102, 103, 104, 105, 108
extracts, vii, x, 57, 68, 127, 136, 138, 153

Index

F

fat, ix, 28, 32, 42, 43, 47, 61, 63
fatty acids, 42, 43, 44, 105
feed additives, 115, 131
fermentation, ix, 33, 68, 106, 131
flavonoids, 3, 138, 147, 157
flowers, 4, 30, 70, 112, 118
fluid, 10, 11, 24, 131, 145, 155
fluid extract, 10, 24, 131, 145, 155
food, viii, ix, 4, 5, 7, 8, 10, 11, 13, 14, 16, 21, 23, 28, 30, 35, 38, 40, 41, 42, 44, 50, 54, 58, 59, 62, 68, 71, 72, 76, 77, 91, 94, 95, 96, 98, 99, 103, 106, 115, 131
food additive, 94
food additives, 94
food industry, 10, 11, 77, 100, 106
food products, 16, 30, 41
food spoilage, 14
fruits, vi, vii, viii, x, 2, 3, 4, 17, 27, 30, 32, 55, 59, 61, 111, 112, 117, 118, 119, 121, 122, 123, 124, 126, 127, 128, 130, 131, 132, 133, 134, 135, 136, 137, 138, 139, 143, 144, 147, 155, 156, 157
fungi, vii, ix, 33, 68, 70, 105

G

gene expression, 85, 87, 99
gene promoter, 91
gene regulation, 130
gene silencing, 131
genes, ix, 28, 47, 53, 85, 88, 89, 90, 91, 108, 131
genetic engineering, x, 68, 69, 91
genus, 74, 75, 76, 118, 124, 126, 129, 132
glycerol, 44, 76, 80, 81, 82, 96, 100, 102, 105
growth, 6, 7, 35, 52, 65, 72, 82, 83, 84, 85, 86, 87, 88, 95, 96, 104, 115

H

harvesting, 8, 52, 113
health, viii, 2, 4, 8, 13, 14, 16, 18, 22, 27, 29, 41, 60, 62, 68, 94, 95, 96, 127, 135, 140
human body, 13, 28, 70
human development, 34, 61
human health, 3, 7, 10, 14, 71, 103, 115, 128, 137

I

inflammation, 13, 14, 34, 100
ingestion, 44
intestinal absorption, 28, 44, 46, 47, 50, 58, 60, 63
intestine, 46, 47, 49, 115
irradiation, x, 68, 81, 87, 90, 106, 109
isoprene, 28, 113, 114
issues, 6, 14, 63, 106

L

light, viii, ix, x, 2, 3, 7, 14, 15, 28, 29, 33, 39, 52, 68, 70, 72, 74, 81, 84, 86, 87, 88, 90, 105, 106, 113, 130, 131
lipid metabolism, 35
lipid peroxidation, 35, 130
lipids, ix, 28, 44, 47, 59, 75, 76, 77, 99, 100, 102, 104, 106, 109
lipoproteins, 45, 46, 47, 64
liquid chromatography, 134
lutein, v, vii, viii, x, 2, 3, 4, 6, 7, 13, 19, 27, 28, 30, 31, 32, 33, 35, 36, 37, 38, 39, 40, 41, 42, 43, 44, 45, 46, 47, 49, 50, 51, 52, 53, 54, 55, 56, 57, 58, 59, 60, 61, 62, 63, 64, 65, 66, 70, 73, 77, 78, 79, 94, 112, 113, 118, 120, 121, 122, 124, 126, 127, 128, 129, 137

lycopene, x, 3, 4, 5, 6, 28, 55, 70, 73, 76, 79, 90, 91, 94, 97, 103, 109, 112, 113, 118, 119, 122, 124, 126, 128, 130

M

macroalgae, 30, 32, 145, 155
macular degeneration, 29, 34, 54, 55, 56, 61
magnesium, 43, 144, 155
magnetic field, x, 68, 69, 86
matrix, viii, 20, 28, 29, 35, 38, 39, 41, 42, 44, 49, 59, 115
matrix metalloproteinase, 20
media, 80, 82, 84, 96
medical, 14, 60, 127
medicine, 118, 121, 123, 129
medium composition, x, 68, 69
metabolism, 29, 47, 53, 56, 61, 83
micronutrients, 3, 46, 51
microorganism, 69, 71, 86
microorganisms, ix, 4, 67, 71, 72, 74, 79, 84, 86, 87, 88, 89, 90, 91, 95, 97, 104, 131
moisture, 12, 37, 38, 39, 40, 50
moisture content, 38
molasses, 80, 81, 82, 96, 98
molecular biology, 88
molecular mass, 11
molecular structure, 6, 88, 114
molecules, ix, 15, 28, 29, 34, 46, 68, 70, 71, 72, 74, 86, 94, 113, 115
mutagenesis, 89, 90, 91, 109
mutant, 56, 87, 88, 90, 91, 97, 101, 102, 103, 109
mutation, 90, 108

N

nutrients, 11, 29, 119
nutrition, 3, 4, 18, 21, 22, 29, 53, 100, 115

O

optimization, ix, 24, 68, 91, 97, 98, 106
oral bioavailability, v, vii, ix, 28, 35, 42, 47, 49
ornamental plants, 115, 118
oxidation, 13, 15, 34, 57, 70, 72
oxidation products, 57
oxidative damage, 34
oxidative stress, 35, 54, 82, 100, 127
oxygen, vii, viii, 2, 4, 6, 12, 15, 27, 34, 70, 73, 82, 87, 113, 114, 130

P

pH, x, 14, 15, 38, 41, 53, 68, 69, 72, 74, 81, 83, 84, 93
pharmaceutical, 4, 7, 8, 11, 76, 91, 94, 95
phenolic compounds, 127, 134, 138, 147, 157
pigment, ix, 2, 4, 7, 14, 15, 16, 19, 23, 28, 29, 34, 39, 48, 50, 51, 55, 58, 60, 64, 65, 69, 83, 88, 94, 95, 108, 113, 122
pigmentation, vii, 2, 9, 56, 75, 88, 96, 115
plant growth, 3, 131
plants, vii, ix, x, 2, 3, 11, 24, 25, 67, 70, 71, 111, 112, 113, 119, 128, 131, 133
polyphenols, 62, 126, 135, 136, 145, 154
potential applications, x, 19, 68, 69
protection, 6, 7, 13, 54, 71, 87
proteins, ix, 28, 33, 39, 41, 44, 47, 65
pulp, 40, 119, 120, 121, 123, 125, 126, 129, 136
pyrophosphate, 72, 73, 86

R

radicals, vii, 2, 6, 12, 13, 34, 71
reaction center, 113, 114
reactions, 70, 72, 74, 88, 95

reactive oxygen, 13, 33, 71, 114
recovery, x, 68, 69, 92, 93, 104
residues, 82, 95, 96
response, 11, 42, 48, 53, 66, 80, 83, 98
retinol, 44, 48, 71
room temperature, 41, 89, 91, 109
rosaceae, vi, vii, x, 111, 112, 117, 118, 119, 120, 121, 123, 124, 125, 126, 128, 129, 134, 135, 139
rose hips, 118, 132

S

salmon, viii, 2, 4, 7, 95, 115
serum, 28, 34, 48, 56, 57, 58, 63
shrimp, viii, 2, 4, 7, 8, 9, 10, 12
shrubs, 112, 117, 122, 123
species, x, 6, 12, 13, 15, 29, 33, 41, 50, 61, 68, 71, 75, 76, 78, 87, 111, 112, 114, 117, 118, 119, 121, 122, 123, 124, 126, 128, 129, 132, 138, 139, 145, 147, 148, 153, 154, 155, 156
stress, 83, 86, 95, 101, 153
stress response, 101
substrate, 48, 76, 81, 82, 84, 98, 108
sugarcane, 80, 81, 96, 98, 109
synthesis, ix, 68, 77, 78, 80, 82, 84, 85, 87, 88, 103, 107, 113

T

techniques, 9, 11, 40, 69, 88, 90, 91, 92, 94, 103, 131
technology, 93, 146, 147, 148, 156, 157
temperature, x, 11, 15, 39, 53, 63, 68, 69, 70, 72, 74, 83, 84, 85, 97, 101, 106, 109

transport, 41, 43, 44, 46, 47, 55, 61, 62, 65
treatment, 12, 34, 37, 39, 40, 50, 98, 99, 127, 153

V

vascular diseases, 123
vascular endothelial growth factor, 64
vegetables, viii, 4, 27, 30, 32, 38, 40, 43, 50, 55, 56, 59, 61, 63, 70, 115, 130, 135, 137, 138
vitamin A, 3, 6, 7, 16, 55, 70, 71, 73, 99, 115, 131
vitamin C, 134, 135
vitamin E, 63
vitamins, 62, 75, 126

W

wastewater, 80, 81, 95, 98, 100, 105, 106
water, 5, 37, 39, 65, 70, 71, 80, 82, 93

X

xanthophyll, vii, viii, 2, 6, 27, 28, 41, 59

Y

yeast, vii, ix, 68, 69, 70, 73, 74, 75, 76, 77, 78, 80, 81, 82, 83, 84, 85, 86, 87, 89, 90, 91, 92, 93, 95, 96, 97, 98, 99, 100, 101, 102, 103, 104, 105, 107, 109